Date: 2/13/12

STRATEGIC PRODUCT CREATION

STRATEGIC PRODUCT CREATION

DELIVER CUSTOMER SATISFACTION FROM EVERY LEVEL OF YOUR COMPANY

RONALD L. KERBER and TIMOTHY M. LASETER
with Max Russell

McGraw-Hill

New York   Chicago   San Francisco   Lisbon   London
Madrid   Mexico City   Milan   New Delhi   San Juan
Seoul   Singapore   Sydney   Toronto

1 2 3 4 5 6 7 8 9 0 FGR/FGR 0 9 8 7 6

ISBN-13: 978-0-07-148655-2
ISBN-10: 0-07-148655-0

This publication is designed to provide accurate and authoritative information in regard to the subject matter covered. It is sold with the understanding that neither the author nor the publisher is engaged in rendering legal, accounting, or other professional service. If legal advice or other expert assistance is required, the services of a competent professional person should be sought.

> —*From a declaration of principles jointly adopted by a committee of the American Bar Association and a committee of publishers.*

This book is printed on acid-free paper.

McGraw-Hill books are available at special quantity discounts to use as premiums and sales promotions, or for use in corporate training programs. For more information, please write to the Director of Special Sales, Professional Publishing, McGraw-Hill, Two Penn Plaza, New York, NY 10121-2298. Or contact your local bookstore.

CONTENTS

Contents

ACKNOWLEDGEMENTS

Like all effective product creation efforts this book melds the varying perspectives of a wide range of people. Each offered insight that helped us better understand and communicate the message we seek to share. First, we want to thank the executives and practitioners who shared their time and valuable experiences.

Although these individuals are many and are cited throughout the book, we want to give particular thanks to those who supported and facilitated our field research. Jeff Fettig, CEO of Whirlpool Corporation, took a personal interest in our work, and his key executives provided additional insight and reinforced co-author and former Whirlpool executive Ron Kerber's experience in the company's ongoing investment in building an innovation culture.

Don Goodman, President of Disney Imagineering, readily agreed to our interview request and Marilyn Waters worked diligently to arrange our visit and subsequent requests for information and chapter reviews. John Helferich, former Vice President of Research & Development, now Vice President of University Research for Mars Incorporated introduced us into his organization, and Bob Boushell, Innovation Portfolio Director, facilitated our interactions throughout the process from field research to final publication.

John Cassidy now retired but, formerly Chief Technology Officer of United Technologies Corporation gave us access to a wide range of managers across the diverse business units of UTC. Greg Brostowicz, of UTC Corporate Communication, graciously provided some great photos of leading edge UTC products for inclusion in the book. Finally, with the help of Bob Luby, IBM's Vice President, Public Sector SCM, we were able to gain insight into the changes occurring at IBM and its movement toward a customer focused research enabled organization. Grace Lin, an IBM Distinguished Engineer and Global Sense and Respond Leader was invaluable in her persistence to arrange interviews and access to key people

Second, we thank the Batten Institute and the Darden School for providing financial support and the research infrastructure for our work. The Batten Institute, a pre-eminent educator and thought leader of entrepreneurship and innovation located at the University of Virginia's Darden Graduate School of Business, awarded a Batten Fellowship to Ron Kerber and helped fund the research and writing. We hope that this book is in line with the Institute's mission to sponsor projects that illuminate the best practices in the areas of entrepreneurship and innovation.

Third, we would like to give special thanks to Max Russell who tirelessly edited and re-edited our drafts and listened to our often hot debates about the key concepts. His sense of humor and patient demeanor made it possible for two firm-minded (or hard-headed) individuals to collaborate in a way that yields a coherent whole full of breakthrough thinking rather than suboptimal compromises. It was at times painful, but seeing the end product has been enormously rewarding. Although his critical eye dramatically enhanced the writing quality for the entire book, any errors or omissions are exclusively our own responsibility.

Acknowledgements

Finally, we would like to thank our families, especially our wives, Kathy Kerber and Jody Laseter, who often listened to us rant about the frustrating process of writing a book. It takes enormous energy, both mental and physical, and hundreds of hours that could otherwise be spent in the more enjoyable pursuits of family life in the lovely rural community of Charlottesville, Virginia.

Ronald L. Kerber
Timothy M. Laseter
Charlottesville, Virginia
August 2006

DEDICATIONS

I dedicate this book to my children, John Andreas, Mark Densel, Stephen Louis and Jacqueline Kelly who have brought me great joy and pride.

Ron Kerber

To Jody, Josh and Cecilia. Thanks again for allowing me the luxury to pursue my passions.

Tim Laseter

INTRODUCTION

Revenue growth in your company will only occur when you create and sell something that your customers want. Continued growth only occurs when this happens consistently through an institutionalized process that captures innovation and customer insight and integrates those concepts into the culture of the company. This book captures the experience of corporate leaders who have successfully done just that. Great leaders achieve the appropriate level of confidence to be successful through an ongoing process of reflection and continuous investment in learning.

Reflect on your company's performance in strategic product creation. Are you satisfied with:

- The contribution of new products or services to your revenue?

- Your product position relative to the competition?

- Your speed to market with new products?

- The stream of innovative ideas feeding new product creation?

- The success rate of converting those ideas into real products?

- The effectiveness of your product creation organization?

If you answered "no" to any of these questions, some—perhaps all—of the concepts and examples provided in this book can be of value to you. Although it draws heavily on the experiences of the authors in product-based businesses, our research has validated the broad applicability of the concepts. The guidance can apply to a product business, a service business, or the service component of a product business. While the term "product creation" is used most often, the concepts and principles presented here can work for any company that wants to create products or services that respond to customer needs.

Overview

* ✱ *

In the following chapters, we offer proven techniques for structuring, staffing, and managing the product creation process. The book presents ways to create a corporate environment that fosters innovation with the discipline necessary to turn great ideas into profitable products. It explores how to integrate research and development and product creation projects into a competitive product portfolio and how to manage business processes to support product creation. Finally, it shows how to capture the full capabilities of suppliers by also viewing them as a talent pool for process improvement and product innovation.

The first chapter sets the foundation for the remainder of the book by providing an overarching framework for integrating product creation deeply into the fiber of an organization and offering a set of key tenets of effective product development.

Part I-Key Concepts outlines the key issues to consider in developing the people, processes, and practices for effective product creation. **Part II-Case Studies** captures the transformation experiences of well-known, successful companies that have approached product creation in distinctly different ways. At the same time, they also have many common threads worth noting.

Throughout, and especially in the case-study chapters and real-world examples, this book offers you insight into where your product creation process stands today relative to best practices. Such insight can help you decide where you want to take your company in terms of product creation, what capabilities you need to develop and embed in your organization, and how to set priorities on what you need to do to achieve your vision.

If you already have a relatively effective product creation capability, consider this a handbook to help refine it and to supplement your training program. If, on the other hand, you do not have a well-defined product creation capability, consider this a guidebook for creating one.

Chapter 1, **Product Creation Leadership**, explains how product creation provides the lifeblood for any business and highlights the critical role of senior management in building and sustaining that capability. It then presents a framework for integrating three portfolios necessary for continuous business renewal: the product portfolio, the portfolio of advanced technology capabilities, and the portfolio of product creation projects. Finally, the chapter highlights a set of seven principles of effective product creation that can be found woven throughout the remaining chapters.

Chapter 2, **Product Idea Generation**, examines how to understand your marketplace by observing customers through various lenses in natural and managed settings, looking outside your industry for new ideas and trends, and listening carefully to customer feedback from as many sources as you have available. By doing so, you can capture the voice of the customer, understand it, and integrate that knowledge into new product and service opportunities.

Chapter 3, **Technology Capabilities**, explores the role of the research organization in supporting product creation. Although advanced research must be kept distinctly separate from product creation efforts because of the inherent uncertainty of invention and the need for discipline in developing products on a timeline, it must be market focused. Developing a suite of technology capabilities to support new products offers the opportunity to be the game-changer rather than merely a fast-follower of others' innovations.

Chapter 4, **Product Creation Process Design**, examines in detail how to develop a stage-gate[1] product creation process, the key principles for managing it, and how to integrate the process into corporate planning. Examples of key questions and milestone decision issues drawn from personal experience accompany descriptions of each stage of the process. This chapter also addresses

---

[1] Stage-Gate® is a registered trademark of Product Development Institute Inc. The term is used here in its generic sense to denote a phased product creation process that includes stages and tollgate reviews.

the importance of functional support to the process, how to ensure it, and associated dynamic staffing issues common to product creation projects. Woven throughout is the importance of senior management in guiding, staffing, and supporting the process.

Chapter 5, **The Competitive Product Portfolio**, explores the management of the company's integrated product strategy, which combines customer understanding and external market trends to stimulate new product ideas. In addition to the process concepts, the chapter also covers product architecture attributes that facilitate a robust product portfolio without excessive cost of complexity by focusing on product platforms, modular design principles, block upgrades, and the concept of "the green line."

Chapter 6, **People and Project Management**, describes the critical, softer side of building a product creation capability. It offers insights into the characteristics of various organization structures that support product creation as well as guidance on how to motivate and manage people within them. The chapter also addresses project management principles and how they apply to the product creation process including the very effective but under-utilized concept of red teams. Finally, it discusses management discipline, project initiation, dealing with projects of various sizes and scope, staffing issues, management tools, process documentation, and continuous process improvement.

Chapter 7, **Innovation Partnerships**, examines how to integrate and leverage suppliers and external organizations into the product creation process. It highlights the need for a good cultural fit with partners and the various roles suppliers can play depending on the characteristics of their products. It also tackles the sensitive issues of intellectual property rights and branding.

Chapter 8, **Walt Disney® Imagineering**, examines Disney's world renowned think tank and its product creation process for attractions at parks like Disneyland, Disney World, and Epcot Center through the eyes of its senior management team.

Chapter 9, **Mars Incorporated**, features an organization with category-leading brands instantly recognizable throughout the world: snacks including M&M'S®, Snickers®, Dove®, Starburst®, and Skittles®; pet food brands such as Pedigree®, Whiskas®, and Cesar®; and "main meal" products under the Uncle Ben's® brand. Despite the company's success, Mars Incorporated has managed to maintain a rather low profile as a privately held company.

Chapter 10, **Whirlpool®**, describes the innovation history of the world's number one manufacturer of home appliances. Since the 1980s, Whirlpool has grown from a regionally-based manufacturer to a global powerhouse by transforming and continuously improving its product creation process.

Chapter 11, **United Technologies Corporation (UTC)**, explores product creation at this $43 billion industrial giant. Although low on the name-recognition horizon to the average consumer, the company's Otis elevator group leads the global market in elevators, escalators and people movers. Its Carrier brand is the worldwide market leader in heating, ventilation, and air conditioning systems for residential, commercial, industrial and transportation customers. And UTC's Pratt & Whitney, Sikorsky, and Hamilton Sundstrand units supply aircraft engines and major aerospace systems for most of the world's commercial and military aircraft industry.

Chapter 12, **IBM**, describes how the company has used product creation to reinvent itself several times evolving from computing scales to tabulators to work-time recorders to general purpose business computers to super computers to personal computers to software and business services. The corporate icon and role model of the 60's and 70's, IBM lost its way and needed the turnaround leadership of Louis Gerstner and his relentless focus on customer needs to take the company back to a position of strength.

Make It Your Own

* ✳ *

Although *Strategic Product Creation* will provide insight for improving any company's product creation process, it is a handbook or guidebook—not a cookbook. Draw your lessons and tailor them to fit your own industry, business model, corporate culture, and business objectives.

Effective leaders of product creation foster innovation. They also exercise discipline. They articulate a product vision for the company and set the boundaries for the product creation teams. They build an environment that inspires creativity and innovative thinking. At the same time, they monitor progress, exercise judgment, and make informed decisions based on the business case for each new product or idea. They also ensure that not only the goals and objectives for new products, but also the rationale for decisions made during the process are well understood and accepted by all.

If you seek to achieve these aims, you will face some critical decisions:

- What product creation outcomes do I want?

- Who should lead needed changes in my product creation process?

- How can I embed a new or revitalized process in the organization?

- How can I make the process enduring and relevant over the long term?

- Do I need to change my organization structure to make that all happen?

- Do I have the right people?

Taken individually, the ideas in this book may not seem unique. However, applying them together, and with commitment, can produce remarkable results. The guidance and case studies make crystal clear that every company needs and can develop a product creation capability to compete more effectively.

If you care about the ongoing success of your business, you need to do whatever you can to foster the development of a capability for strategic product creation. Although this book does not tell you *what to think*…it will focus your attention on what you should *think about*. Serious reflection on the strengths and weaknesses of your organization offers the first step; failing to do so, may be the final step towards extinction in today's global business world.

Part I
KEY CONCEPTS

Chapter 1

PRODUCT CREATION LEADERSHIP

———— ✳ ✳ ✳ ————

Delivering Business Results

Business starts and ends with the customers. As William Dillard, who founded upscale Dillard's Department Store in 1938, explained, "You can't sell from an empty wagon." Effective corporate product creation requires more than the newest, hot technology or accomplishing what others said could not be done. It is about making products that customers will buy at a price that exceeds cost. Period.

That assertion may sound simple, or even simplistic. It is neither. Without the right kind of product creation leadership and the right innovation environment, a company may fall into the trap of focusing exclusively on the next new thing, embracing technology for technology's sake, or obsessing on delighting the customer, yet failing to do any or all of that at a profit.

Creating Products for Competitive Survival

* ❊ *

Truly vibrant companies continuously create new products to drive revenue, profitability, and growth. Corporate executives take many different approaches to sustaining successful business leadership. Some adopt a business model based on product innovation. Others ground their businesses in operational excellence, cost leadership, or customer service.

Examples of successful companies built on each of these models abound. And although operational excellence, cost leadership, and customer service can all deliver a formidable advantage, customers will only buy products or services that suit them best. Simply put, low cost, high efficiency, or exemplary service alone are not sustainable strategies without competitive products.

To sustain leadership, companies must have a competitive product portfolio achieved by an ongoing process for creating new products and services. This does not mean that a company must have the most innovative products or come first to market with new ideas. It does mean, however, that a company must keep its product portfolio fresh and competitive through some sort of product creation or renewal process, even if that means being a fast follower or an acquirer of new offerings.

Process is the operative word here. Success ultimately depends on a company's ability to sustain a competitive product or service portfolio through a product creation process defined by disciplined creativity and guided by an integrated product strategy.

4

Integrating Product Creation Strategy

* ** *

Critical to creating and sustaining a competitive, profitable business model is integrating the complex array of processes and capabilities that influence the way a company adds new products to its marketplace portfolio. As illustrated in Figure 1.1, building a competitive product portfolio involves the integration of many different organizational activities each with its own purpose and timing. The linkages among technology capabilities, product creation projects, and the product portfolio in the market serve as the core of an integrated product creation strategy. These linkages resurface in more detail as this book explores the intricacies of product creation. But those core

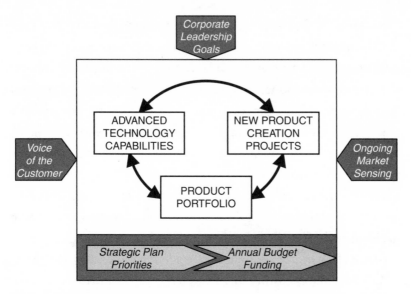

Figure 1.1 Building a Competitive Product Portfolio

5

activities must also operate within the context defined by broader business processes including corporate leadership goals, ongoing customer understanding and market sensing activities, the strategic planning process, and the annual budgeting process. The responsibility for all of these varied linkages falls on the shoulders of corporate leadership. Senior management must coordinate a vast array of moving parts to ensure sustainable, ongoing product creation and its ultimate output, a competitive product portfolio.

Executive Leadership and Accountability

* ✳ *

Despite the critical importance of new products, amazingly, many companies lack the ability to develop new products and services consistently and efficiently. Accountability for this fundamental shortcoming rests squarely at the feet of executive leadership. And so does responsibility for correcting it. Effective corporate product creation requires active executive leadership to ensure a constant focus on delivering business results through profitable products that delight customers. Executive leaders need to inspire—and channel—the creativity of the organization while maintaining an appropriate level of discipline. Leaders must make the strategic decisions: where to invest, when, and how much—and the most difficult decision of all, when to walk away. Getting the latter right can make the difference between product portfolio success and failure.

The most effective role for senior management, however, involves building the culture, the processes, and organization to create a sustainable *capability* for corporate product creation. Direct, hands-on management of innovation activities may keep the organization on course in the short term, but only a culture that supports a well-developed capability for creating new products and services can sustain competitive advantage.

Integrate three portfolios. A successful product creation capability generates a competitive product portfolio through the ongoing management and integration of two other portfolios:

- The portfolio of new product and service creation projects

- The portfolio of technology capabilities

The portfolio of new product and service creation projects directly links to the product portfolio through a disciplined product creation process and product introduction plans. In turn, a portfolio of technology capabilities connects to the project portfolio and ultimately to the market through the use of technology roadmaps (more on that later). And the product portfolio, in turn, influences the content of the portfolio of technology capabilities— completing the cycle. Senior management plays a leadership role in keeping the organization focused on the customer through the ongoing nurturing and culling of all three portfolios.

Although all three portfolios must stay market-driven, each plays a different role in the ongoing process of innovation, and each requires a somewhat different management mindset.

The product portfolio needs constant updating—through a mix of major introductions and minor upgrades—to outdistance competitors and maintain customer interest. Although the minor upgrades do not require as many resources or as much rigor as the formal product creation process, they are enabled by a common planning process and product architecture.

The portfolio of product creation projects provides the sustenance for truly exciting renewal of the product portfolio. As such it should reflect a mix of new-to-the-world products, major upgrades, and product line extensions. The appropriate mix will vary by business of course, but a well-managed, diversified project portfolio reduces the impact on the company of the failure of any individual product creation project.

A portfolio of technology capabilities nourishes the project portfolio, and at times, directly feeds the product portfolio. Technology capabilities need not require breakthrough engineering. They may simply reflect new applications from another industry. Management cannot control the portfolio of technology capabilities with the same precision as it does product introduction plans and product creation projects. Nonetheless, the technology capabilities portfolio can be managed with the same discipline and accountability as the other two. Again, common sense applies. Technologists identify emerging capabilities for consideration in the portfolio, and senior management examines each with a focus on the marketability of the technology. Senior management may also set goals for the technologist to develop new capabilities based on gaps in the desired product portfolio.

Adhere to key principles. Managing the three interrelated portfolios—the advanced technology capabilities, product creation projects, and the product portfolio—provides the core of a corporate product creation capability. Creating a true competitive advantage from that capability requires adherence to a number of principles:

- Integrate throughout

- Look outside

- Don't reinvent

- Seek disciplined speed

- Staff right, staff light

- Succeed by acknowledging failure

- Mindfully manage tradeoffs

These principles create a set of interwoven themes that appear and reappear in various topics throughout this book.

Integrate throughout. Although the integration of the three portfolios serves as the heart of the corporate product creation capability, this process must align with all core business processes to deliver the best results. First and foremost, top-down corporate goals set the broad expectations for the organization. Ongoing market sensing efforts that capture the voice of the customer and external business intelligence offer insight to anticipate market needs. Equally important, strategic planning and annual budgeting prioritize and constrain the product portfolio. No company can afford to explore all potential market needs. The essence of strategy involves deciding what not to do as much as it involves deciding what to do.

Integration goes beyond the business processes to include the people in divisions and functions throughout the organization. Senior management should constantly seek synergies between business units and eliminate redundancies—while avoiding undermining business unit accountability. A global, integrated product vision can engage and even inspire the workforce at all levels as it points the way towards a clear future for the company. Buy-in to the vision of future product and service creation helps ensure essential functional support at critical junctures of product creation as well as in the more transactional and often contentious budgeting and staffing processes.

Very few companies have a completely integrated capability. Even those that do must constantly strive to maintain and improve it. Product creation is a dynamic capability requiring care and feeding to sustain it over time. Left alone, it will atrophy and die. This constant emphasis on improvement leads to the next recurring theme of this book, look outside.

Look outside. A focus on internal integration throughout could unintentionally lead to a perilous, insular view. A company must look outside to understand how customers behave in their own environments. But that is only the start. Connecting with the external market protects product developers from the common affliction of becoming too internally focused and technology-driven. An outside view also exposes product developers and technologists to other industries offering new ideas and capabilities that could be adapted for new products, functions, or features. Looking outside to the market tests whether a technology addresses a real customer need.

The directive to look outside applies beyond the technology experts however. Studying competitors to understand their strategic direction proves integral to forming one's own product strategy. Such study requires more than a review of a competitor's product offerings, product introduction announcements, and formal statements of strategic intent. True insight means understanding the competitor's values and management tendencies and anticipating their actions.

Much less threatening, outside suppliers and other partners can open many doors to new product and development opportunities. Working with these so-called outsiders can create strategic relationships that give both parties a business advantage. Further removed from suppliers and partners, looking outside at the broader landscape can help spot shifting trends in such dimensions as color, industrial design, new materials, new flavors, or new dynamics in popular culture and lifestyles.

Finally, companies should look beyond their own industries for new business processes and best practices that may apply to their products and services. Business-changing paradigms often come from the outside. Taichi Ohno, father of the famous Toyota production system, took inspiration for his innovative ideas from U.S. grocery stores as well as the original principles of Henry Ford's assembly line (something that Ford "borrowed" from the meat packing industry at the turn of the 20th century).

Looking outside provides market guidance and helps avoid being blind-sided by the competition or other external forces. Careful surveillance can help a company become a game changer rather than a sideline victim. Senior management must constantly encourage an external focus. Most companies don't do it naturally. The time taken to do so, however, can prove well worth the extra effort required.

Don't reinvent. Maintaining a healthy, open outside view can also protect an organization from the often treacherous "not invented here" mindset. This feeds directly into the next guiding principle of product creation: don't reinvent the wheel. Given the relentless pressure on resource allocation, companies simply cannot afford constant reinvention in the quest for an innovative product portfolio. Uncovering existing solutions to a current technology challenge allows researchers and innovators to focus on the truly unique problems that can produce a competitive breakthrough.

Many, if not most, new ideas, technologies, and business processes reflect adaptation of concepts previously applied in other contexts rather than fundamentally original innovation. Apple's wildly successful iPod and iTunes music retail platform integrates components and concepts already found in existing consumer products and websites, but in a creative way that better meets consumer needs. Product creation leaders like Procter & Gamble look first for existing solutions to meet consumer needs. Such a philosophy led to the Crest® SpinBrush™ and the Swiffer™ dust-mop.

Reinvention creeps up internally as well. Without thoughtful planning of the product portfolio and the supporting product architecture, companies may needlessly reinvent portions of their own product portfolios. To minimize wasteful reinvention, explicitly design for reuse. A well articulated product platform strategy allows a company to extend a reusable platform across multiple brands and geographic regions. Using a modular design strategy for products allows the use of components and subsystems across a variety of products without a total redesign for each iteration. Because reuse may require some adaptation from application to application, careful planning and product creation experience become even more important in finding the right balance.

The "re-use don't reinvent" mantra also applies to the processes and systems for product creation. Many good models for product creation exist already. The case study chapters in this book featuring Disney Imagineering, Mars Incorporated, Whirlpool, United Technologies Corporation, and IBM document a variety of approaches. Although each company must ultimately make the new product creation process its own, existing approaches offer a good place to start.

Seek disciplined speed. Although being fast to market can provide a competitive edge, introducing a product prematurely can lead to quality problems, product failures and significant damage to the brand. Senior management must apply experienced judgment to maintain the correct balance between discipline and speed.

Faster product creation efforts typically require fewer resources, but only up to a point. For example, in an industrial application, meeting a faster project timeline might require buying more expensive tooling or reworking the tooling to accommodate later design changes. A slower project timeline avoids the cost of expediting premiums and minimizes the risk of rework, but the basic resources stay in place longer incurring expenses during the extended timeframe. As the insightful Murphy asserts: work will expand to consume the time available.

Although faster to market can provide an advantage over slower competitors, too short a timeline may force tradeoffs in product features that make it less attractive to the customer. Even if the shorter timeline doesn't force feature tradeoffs, the end product or service may become more costly because of less time to explore the most efficient operational processes. In the worst case, a product introduced before its time could suffer from quality problems, perhaps even prompting a recall. Such an event could erase all profit potential or even result in a loss and permanent damage to the brand. While high speed involves identifiable dangers, the fastest usually win the race.

The timing decision inevitably involves a choice between risk and reward. How big is the competitive threat? How fast does the profit potential decay? How fast can the team push without incurring excessive premiums or introducing quality problems? The answers to these questions define the optimal pace for disciplined speed. The best solution typically pushes for launch as quickly as prudent with a plan to improve both the product and cost with future product upgrades.

Beyond the specific project, senior management must also think about continuously improving the company's product creation capability. Investing in making the product creation capability stronger increases the ability to operate safely at a faster pace. An old clunker of an automobile will careen out of control if driven too fast. A high-performance race car can remain safe and in control at enormously fast speeds. Yet, even the race car can spin out of control in the wrong hands or when pushed beyond its capabilities.

Staff right, staff light. Extending the high-performance race car metaphor to product creation, success lies largely in the hands of the driver with unswerving support from everyone from the pit crew to the mechanics to the sponsors. The winning team attracts the best people. Attracting the best also demonstrates a company's commitment to new product creation. When truly committed to innovation, a company's functional career ladders to senior positions should all include a few product creation "rungs" to scale.

With the best people fully committed to product creation, a company can staff light. The most critical success factor for a product creation project is staffing right. Even though outstanding people earn higher wages, fewer high performers deliver much more. Average performers produce average results and invariably consume a disproportionate amount of senior management time and attention. The old maxim, "you get what you pay for," has never rung truer.

To transform a mediocre product creation organization, cut the mediocre staff until it hurts. And then cut some more. Once the underperformers have departed, reinforce the critical areas that need resources the most. If the task involves building a new capability, set wage structures to attract the very best. Hand-pick and nurture them. A small contingent of passionately committed individuals can deliver extraordinary results.

Succeed by acknowledging failure. Failure is painful. Especially for passionate, bright people—the kind of people that companies want in their product creation organizations. Ironically, innovation is inherently fraught with uncertainty and risk, which inevitably leads to failure at times. Acknowledging failure without recrimination allows individuals and the organization to move forward and to pursue more productive efforts. The ability to recognize dead-ends, stop the effort, and move on requires great judgment and self-confidence. Team members need to know that management will not shoot the messenger who identifies a dead-end project.

Senior management must encourage a culture that demands excellence, but accepts failure as a natural outcome of pursuing excellence aggressively. If staff members fear failure, they will intentionally or unintentionally hide it from management. They will bury risks in footnotes and appendices rather than raising them for healthy discussion and debate. Far better is to get the issues on the table before they become program-threatening crises. If management does not respond constructively to bad news, even the best team members will avoid highlighting their challenges, believing—more often hoping—that they can solve them alone.

In the end, the success of any product creation process rests on the foundation of experience and good judgment built by senior management. Experiences—successes and failures alike—build the capacity for good judgment. Without the occasional failure, a company has probably not pushed aggressively enough. Failing without understanding why, however, misses the opportunity to instill sound judgment for future decisions. Failures can go two ways: they can be a complete loss or a valuable investment in learning.

Manage the tradeoffs, mindfully. As evident in the discussion of each of the preceding principles, senior management must balance many counteracting forces in developing the capability for creating new products and services. A desire for well thought out internal integration cannot blind the organization to the dynamics of the external market. And though the external market exhibits a constant state of flux, companies cannot afford to continuously chase the newest trend. Mindfully responding to the market within the context of a stable, long-term plan avoids needless reinvention. Appropriately balancing risk and reward ensures that the organization moves at a disciplined speed without careening out of control. And although no organizational capability depends upon people more than product creation, too many people can hamper success. Staffing right requires staffing light, which implies using the very best people available. Even though every project launched must have the potential for success, some must fail. A lack of failure reveals a fear of risk—a sure road to mediocrity.

All of these difficult tradeoffs require strong leadership, experienced judgment and clear decisions to set the expectations for the organization as a whole and for individual projects as well. Examples include setting stretch goals for a particular technology area that could lead to a product breakthrough, shortening the typical time to market goal of a project by 30 percent, or allowing for a longer development cycle time to assure that aggressive cost and quality goals are met. Such tradeoffs do not come easy and most require constant readjustment to attain the appropriate balance in light of ever changing circumstances. Risk and reward. Short-term versus long-term. Consistency with flexibility. Haste not waste. Fundamentally, senior management must mindfully manage tradeoffs to build and maintain a product creation capability that provides a competitive advantage.

Chapter 2

PRODUCT IDEA GENERATION

———— ✳ ✳ ✳ ————

Sensing Customer Needs and Market Opportunities

Executive leadership sets the tone and timbre for generating innovative product concepts. At their best, executive leaders inspire the organization to reach new product creation heights. Steve Jobs, founder and CEO of Apple®, continues to receive accolades for his technological vision and his goal of "making a dent in the universe." The original Apple computer brought computing to the average consumer. Its aesthetically pleasing and functional design became Apple's enduring hallmark.

Decades after the company's first breakthroughs, the runaway success of the Apple iPod® only serves to reinforce Jobs' original ideals. But top-down ideation from the executive suite does not always prove so fruitful—even with Apple's talented workforce. The Apple Newton™, for example, serves as a lesson in technology-driven as opposed to consumer-driven product creation.

John Sculley, Apple's CEO from 1985 to 1993, coined the term "personal digital assistant" in the late 1980s and charged a product development team to bring the concept to life. The Apple Newton debuted in 1993 at a price of $699. Developed by technologists without the aid of the marketing staff, the bulky, unreliable device proved a debacle that nearly tanked the entire concept of a PDA.

Learning from Apple's mistakes, Palm Computing® introduced the Palm™ Pilot in 1996. Priced at $499, the Palm Pilot downplayed the technological "wow" factors and focused instead on the more mundane but critical tasks that consumers needed from a portable device—scheduling, personal contacts, to-do lists, and a simple interface with a personal computer. By capturing customers' real needs, Palm captured the market. The battle, however, may not be over. The Apple iPod line now offers the two core functions of a PDA—scheduling and personal contact files—that appeal most to a new generation of consumers increasingly addicted to such powerful devices.

In most cases executive management plays a less active role than Steve Jobs did in generating product ideas. More commonly, executive leadership drives product creation by setting growth targets that require new products, services, or even entirely new businesses. Organizations well-primed through ongoing market sensing efforts can quickly respond to such challenges by systematically generating new product and service ideas, sensing customer opportunities that feed innovation, and using a well honed arsenal of tools for testing and refining their ideas. They also understand the importance of reaching beyond customers to assess both positive and negative influences from current and potential competitors, channel partners, and others offering complementary products or services.

Exploring the Full Range of Possibilities

* ✳ *

Groups tend to generate more creative ideas than individuals working in isolation, especially when group members have widely varying experiences, capabilities, and perspectives. Innovation cells provide an effective means for addressing a top-down growth mandate by tapping into this group dynamic. The concept is straightforward but rich in creative potential. As innovation cell members brainstorm product and business concepts, they try to break traditional paradigms and explore the widest range of possibilities. As shown in Figure 2.1 members of an innovation cell cycle through a pattern of expanding and contracting those concepts and possibilities until a well defined, preferred option emerges. In the discovery phase the team may be prompted through a set of exercises designed to challenge existing paradigms—in this case, discontinuities, core competencies, customer experience, economic engine, and orthodoxies.

The discontinuities exercise, shown in the "discovery" cycle of Figure 2.1, challenges the team to identify drastic changes that could fundamentally alter industry dynamics. Continuing with that segment of the figure, the team can also consider how to extend the company's core competencies into new markets. By looking at the customer experience, the innovation cell seeks to find unfulfilled needs. Understanding the economic engine surrounding a current product or service helps to identify untapped profit pools. Finally, this team also challenges existing orthodoxies by asking the provocative question, "What can we do today that would fundamentally alter the business if we could achieve it?"

21

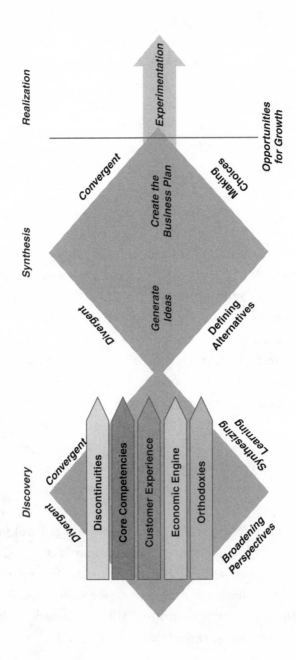

Figure 2.1 The Innovation Cell Concept

After identifying opportunities, the cell then narrows its list of concepts to the best options to pursue.

In the "synthesis" cycle, the team again diverges from "selecting options" to expansively considering ideas that address identified opportunities. For example, having identified an underserved, price-conscious market segment, the team might consider a variety of ways to tap that potential. The segment might be served with a new, low-priced product, but the company might also support that same segment by offering low-cost rentals or even previously owned versions of existing products. After exploring a variety of possible solutions, the team must again converge on a specific business proposition before moving to the "realization" phase of the process where experimentation validates or highlights issues with the initial plan.

Pushing the limits of concepts and possibilities helps to develop a vision of the ideal product. Even though the ideal may not be achievable, such dynamic exploration can energize other efforts that are within the realm of the possible. For example, the Pentagon's vision of an invisible aircraft in the 1960s ultimately led to the stealth plane concept. Though not truly invisible, the technology that resulted from the original vision makes the stealth bomber extremely difficult to detect by standard technologies. More recently, the U.S. Department of Defense articulated a vision of "no safe place to hide," which led to the development of precision guided weapons that have proven critical in the war on terrorism.

Providing Market Sensing Activities

* ※ *

For an innovation cell to produce market-relevant ideas, the members must have a good understanding of customer needs and the marketplace. Ongoing market sensing activities provide that insight. These activities offer a variety of ways for viewing the customer and the competition, which serve as a guide for generating new product concepts and as a touchstone throughout the product creation process. The "voice of the customer" concept has emerged as the most common term for these efforts. In its fullest sense, listening to the voice of the customers means understanding what customers want, need, and will buy even if they have not yet consciously articulated those wants, needs, and purchase intentions themselves.

Voice of the customer entails far more than simply listening. Effective innovation involves all of the senses. The more expansive term, market sensing, augments the idea of satisfying core customer needs with the idea that a product must also be an attractive business proposition, likely to deliver profitable returns.

There is no set formula for market sensing activities. Market sensing should be a broad and ongoing process aimed at identifying an as yet unmet customer need. At the same time, market sensing tools and techniques can also focus more narrowly on a specific project or product in response to an already identified need.

Many of the tools and techniques that have proven most useful to companies in understanding the customer and identifying new business opportunities are described here. All serve as good sources of input to the product creation process. None of them are free of shortcomings, nor do any of them eliminate the need for or diminish the value of sound judgment and experience in determining which features and benefits customers desire most and will actually purchase.

The importance of making customer understanding the basis for generating product concepts cannot be overstated. Nonetheless, companies must take great care not to become mired in the "squishiness" of these processes. While companies must listen closely and define concepts well, they must also zero in on the concepts that show the greatest promise and potential as soon as possible. Some refer to these earliest processes as the "fuzzy front-end." While the fuzzy front-end is vitally important to product creation, the tendency to languish there often delays the formal initiation of a product creation project, which in turn lowers the potential for realizing the maximum return on the product. Markets can change. A competitor can preempt with a product that devalues a new concept. Anything, in fact, that delays time to market or shortens the useful life of a new product can diminish its revenue potential.

With ongoing customer-focused, market sensing activities—operating independently of formal innovation cells or product creation projects—ideas emerge and percolate without necessarily coming to a rolling boil. Overcoming the inertia that can accompany the fuzzy front-end and creating forward momentum often require senior management intervention—a voice from above, informed by experience and sound business judgment. An authoritative voice that says, "Stay here no longer than you must. Listen, look, decide, move on. Your ideas are perishable. They are relevant to a particular place in time and specific to a certain competitive environment. Convert your ideas into reality. Do it expeditiously and well."

Understanding Customer Needs

* ✳ *

The term, "voice of the customer" may evoke thoughts of consumer product companies; the virtues of listening to the voice of
the customer apply equally well to service industries, the business-
to-business world, and even entertainment. Disney's Imagineers
(see Chapter 8), for example, regularly observe guests at Disney's
theme parks for insight into how to improve the guest experience.
They observe firsthand how guests behave while waiting in line
for an attraction and then use those observations to find new ways
to make waiting less stressful, even enjoyable if possible.

On the industrial front, Dow Chemical Company has instituted an "invent to order" approach that starts with understanding its industrial customers, for example, the needs, wants, and
preferences of carpet makers and apparel producers.[1] Through its
efforts to understand the customer better, Dow identified six
unmet opportunities, including a need for a soft stretch fiber that
could stand the heat and chemicals used in manufacturing and
post-sale cleaning processes. Launched in September 2002, Dow
expects its innovation that addresses this need, the XLA™ fiber,
to generate $300 million in sales over 10 years.

Companies can use myriad techniques to try to understand
more about customers and their needs. Every technique has its
own strengths in helping a company determine what the customer
wants, and more important, will purchase. None are perfect.

---

[1] "Inventing to Order: Dow cuts R&D risk by finding out first what its
customers need", Louis Lavelle, *Business Week*, July 5, 2004.

OBSERVATIONAL TECHNIQUES		CONCEPT TESTING TECHNIQUES
o Customer Focus Groups	o Process Mapping	o Concept Cars
o Natural Observation	o Customer Support Monitoring	o Prototyping
o Photo Journals	o Service Support Monitoring	o Usability Laboratories
o Ethnography	o Interactive Internet Feedback	o Product Use Simulations

Figure 2.2 Customer Understanding Techniques

Think of them as lenses. Each lens gives a different perspective of customer needs, and in general, the more lenses used the better the understanding gained. Because successfully sensing the marketplace opportunities requires a variety of lenses combined with sound judgment and experience, the never-ending challenge is to define ever more lenses. Figure 2.2 shows many of the most useful in practice today, each of which will be discussed more fully.

Observational Techniques

*** * ***

The techniques on the left-hand side of Figure 2.2 can be used as ongoing processes or in directed efforts after initiating a formal product creation project. Techniques on the right-hand side, however, require a concept example with which customers or others can interact to capture their perceptions and opinions. Both categories of techniques offer valuable insight for generating and refining product and service ideas.

Customer focus groups. Current customers represent the most common and important starting point for capturing the customer's voice. Focus group sessions have become one of the most popular techniques for doing so. Dating back to the 1930s, the first use of a focus group was to study loyalty and morale among American soldiers.[2] Today, nearly all consumer goods manufacturers and media companies conduct focus groups with existing customers segmented by various criteria, such as brand preference, lifestyle characteristics, age, ethnic groups, or cultural affinity. According to Ann Dahmer, co-founder and president of Smith-Dahmer and Associates and an expert on consumer focused research, certain of these criteria can change over time. Dahmer notes, however, that two of the most enduring segmentation criteria are values and attitudes.[3]

During a focus group session, participants respond to various product concepts presented in a controlled environment. Common practice is not to disclose the identity of the company or the brand under consideration. While the process may start with current customers, involving potential customers can open up product innovation possibilities that would appeal to new customers as well as unforeseen opportunities that would help retain existing ones. This process is greatly enhanced by determining the logical segments of both current and potentially new customers.

Whirlpool conducts frequent focus groups across most of its product categories. Typically, a group of current users, segmented by lifestyle, gather in a room with several versions of a product. A moderator explains the benefits and features of each version.

---

[2] http://www.amonline.net.au/amarc/pdf/research/focusgps.pdf.
[3] Ronald Kerber interview with Ann Dahmer.

Each member of the focus group then rates the product, and the group members collectively discuss their observations while members of the marketing organization observe through a one-way mirror. In one such group, homemakers provided insight that led to the development of a waist-high pedestal for the front-loading Duet® clothes washer.

While focus groups provide useful insight into gauging customer priorities or preferences, what the participants say in a focus group setting does not always accurately predict their actual behavior in the marketplace. A prominent food and beverage company found that consumers in focus groups said that they wanted healthier snack food choices. In practice, however, they were unwilling to compromise on the taste of their potato chips in exchange for less fat or fewer calories. Often customers offer strong support for a product in a focus group, then fail to "vote with their dollars" once the product reaches the market.

Many companies have found focus groups particularly unreliable as a tool for setting specific price points. While focus groups provide depth and emotional insight, they offer relatively little value for rigorous quantitative assessment. Ann Dahmer notes, however, that when conducted with structured statistical techniques, like conjoint analysis, focus groups can provide clear indications of customer preferences for different products or different feature bundles, which may provide insight into acceptable price ranges.

The Honda Element, introduced in 2003, offers another example of the limits of focus groups. According to a January 2003 article in *Forbes*,[4] the Element "was literally designed and sweated

---

[4] Frank, Michael, "Test Drive: 2003 Honda Element EX," Forbes.com, January 6, 2003.

over by the target market—active, suburban twentysomethings. Honda surveyed hundreds of college-age folks and asked what they wanted in a car/SUV and even had them submit drawings and ideas. The Element is what they wanted, right down to the funky shape." Unfortunately, those twentysomethings did not buy the car. A July 2005 *New York Times* article," This Car is for Kids, but Gramps is Driving It," cited a study by CNW Marketing Research that found the average age of Element drivers was around 45 instead of the expected 29.[5]

Rather than a definitive evaluation of specific product concepts, focus groups can provide a richer insight into the customer's product paradigm, that collection of attributes or benefits that the customer expects or would like from the product and its use. Focus groups can also demonstrate the degree to which customers understand the proper use of the product or the scope of its possible uses. Many consumer electronic products—the VCR provides a classic example—are not fully utilized because manufactures paid insufficient attention to ease of use. By defining the product paradigm, a product creation team can often find new ways to communicate product benefits or use, identify additional benefits or uses, or even generate new product concepts.

Natural observation. Observing how customers behave in their natural environments often provides deeper insight into wants and needs than asking for their opinions in isolation. For example, Honda transformed the design of the trunks of its automobiles by observing families in the parking lot of Disneyworld. Company

---

[5] Blumberg, George P. "This Car is for Kids, but Gramps is Driving It," *The New York Times*, July 3, 2005.

engineers roamed the parking lot and saw families struggling to load and unload heavy items, such as baby strollers. A significant hurdle was the relatively flat car trunk lids that put the actual opening several feet above the ground. The engineers realized that a simple bend and extension of the trunk lid downward toward the bumper could reduce the lifting height by a full foot, taking the design back to that of American automobiles of the fifties. Such simple, yet profoundly effective, customer-focused design choices helped propel the Honda Accord to number one in unit sales in the United States in 1989 and made it a 20 time winner of Car and Driver magazine's top ten car award through 2006—more than any other car in the 23 year history of the award.[6]

As noted earlier, Disney Imagineering observes guest behavior throughout its theme parks—how visitors use the various facilities and attractions, what they prefer, what they avoid, and even what might be missing from the experience. The company's product developers, the Imagineers, then use renderings and process maps to solve problems or create new theme park attractions or features. This tradition dates back to Walt Disney's rendering of his original vision of Disneyland (see Figure 2.3). Such meticulous attention to detail helps ensure a consistently high quality experience from the time guests arrive through every ride and attraction and on until Mickey and Minnie bid them good-bye.

Photo journals. Observation does not necessarily require real-time viewing by a third party. Some companies place video cameras in unobtrusive locations to record customer behavior in a natural setting. This has the advantage of minimizing the effect of

---

[6] http://www.hondanews.com/CatID2003?mid=2005121325051&mime=asc.

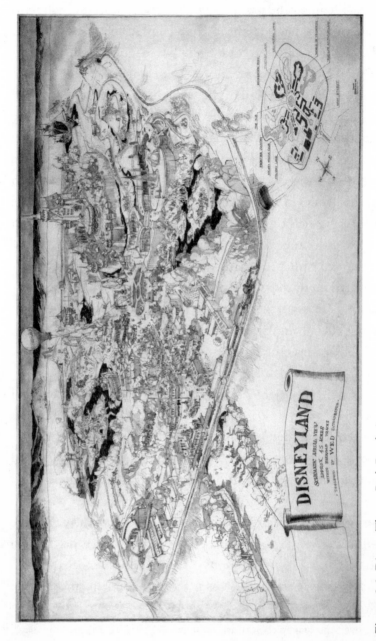

Figure 2.3 Disney Theme Park Rendering

the observer. Just as the Heisenberg uncertainty principle of physics states, an observer unfortunately influences the observed.

Allowing subjects to develop their own photo journals offers another way to minimize the impact of the observer. Masterfoods®, the U.S. division of Mars, Inc., conducted research on teenagers by giving them video cameras to record their own snacking behavior. In another instance, the company gave parents cameras so they could first take pictures of the process they go through to compose an Easter basket and subsequently document the consumption of the goodies by the recipient over the next several days. For its pet food division, Mars asked pet owners to photograph key moments with their dogs and cats, especially feeding. All of these techniques proved extremely useful in analyzing consumer needs and preferences, and ultimately satisfying them.

Although largely qualitative in nature, observation can be structured to extract hard data on preferences as well. Mars observes the choices made by pets when offered two different foods for use in conjoint analysis. In developing its latest pet food innovation, Whole-Meal™, Mars' researchers offered dogs paired options featuring different size, texture, and flavor bones to understand preferences and interaction effects to determine the best product features.

Ethnography. An even more focused iteration of natural observation, ethnographic research produces even deeper insight into the impact of ethnicity, cultural heritage, age, or socio-economic status on consumer needs. For example, in considering further global expansion of its theme parks, the Disney Imagineering group pays particular attention to the "stories" it tells outside of traditional "western" culture. Snow White and Jiminy Cricket won't necessarily resonate in Asian or Latin American cultures.

In like fashion, the company modifies food concessions, including the serving of alcoholic beverages, in its parks depending on local customs and expectations.

Cultural differences represent only one of the many dimensions of interest in ethnography. A virtually endless list of other consumer processes and activities like computer software and hardware use, shopping, transportation, physical exercise, and home furnishings selection can vary dramatically by factors such as age group and gender. Understanding the broader range of consumer behaviors and needs associated with these processes offers enormous opportunities to better serve the full spectrum of the mass market and leads to more informed market segmentation.

Process mapping. Whirlpool takes observation to a somewhat higher level by observing consumers' behavior in their homes and then formally mapping the processes involved in the use of its appliances. Process maps for food preparation, food preservation, kitchen clean-up, and clothes care have revealed that customers often use products much differently than the manufacturer originally intended and that they find ways to compensate for product shortcomings. For example, Whirlpool observed that older consumers often pre-rinse plates before placing them in the dishwasher even though the appliance design provides for thoroughly cleaning a plate containing a complete 12-inch cake.

Process maps have also helped Whirlpool identify opportunities and generate concepts for completely new products. For example, the map of the clothes care process, shown schematically in Figure 2.4, highlights that traditional washers and dryers really only have utility for a portion of the user's overall clothes care needs. Drawing on this insight, Whirlpool developed the Portable Valet™,

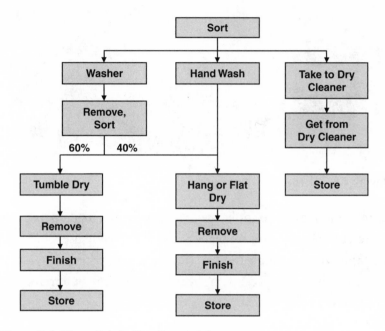

Figure 2.4 Home Fabric Care Process Map

an appliance that refreshes, deodorizes, and de-wrinkles the clothes that the consumer would normally take to a commercial drycleaner. Sometimes, however, a process map leads to a dead end. An affordable appliance that could eliminate ironing, folding, and storing clothes would be a tremendous breakthrough but presents an unattainable goal given existing technology. Identifying a customer need does not necessarily mean that a viable solution exists.

Maps of the food preparation process also stimulated the development of Whirlpool's Polara® Refrigerated Range (see Figure 2.5). Combining the features of two separate appliances with almost diametrically opposed functions, this innovative product keeps a prepared meal, like a casserole or lasagna, refrigerated until a predetermined cooking time when the oven feature

Figure 2.5 Polara Refrigerated Range

comes into play. The appliance as originally designed operates on a preset timer, with future plans to possibly enable users to access the appliance through the Internet to adjust the timing if, for example, the kids' soccer game goes into overtime or a traffic jam delays return from work.

Customer support monitoring. Companies receive direct and indirect feedback from customers continuously and through multiple channels. Few fully leverage this "free" information to generate new concepts or improve the product creation processes. For example, most major corporations have "help lines" to assist customers in understanding and using their products and to solve problems. Few executives or their design teams, however, take time to listen to the calls. If they did, they would know that physically monitoring these customer interactions can give a product creation team enormous insight into not only how the customer understands product features and use, but also what creates customer concerns and dissatisfaction. Information also comes not only through the questions that customers ask, but also those they don't. This is a great place to gain a better understanding of the customer's product paradigms.

Service support monitoring. Call centers represent only one of numerous forms of customer feedback. In many cases companies offer warranties and service contracts for the products they sell. Monitoring warranty and service records and repair parts usage offers a unique opportunity to understand problems with the product, both real and perceived. Automotive manufacturers spend billions each year satisfying warranty claims. Ford Motor Company aggregates warranty information by subsystem and provides the data to the development teams responsible for those systems. When developing new systems, the team must simultaneously focus on new product benefits and lowering warranty costs.

In the case of home appliances, heating and air conditioning units, and elevators, the size of the product and its lack of mobility demand that it be serviced where installed. In such cases, repair personnel in service trucks travel to the use site to solve the problem. Many companies encourage development groups to ride along in the trucks to gain first-hand knowledge of the product environment as well as the problems experienced by customers. Reported problems often stem from a consumer's misunderstanding of how to use the product rather than a faulty product failure. Even then, a product creation team should view such problems as design failures at some level. A truly great product has clear instructions, or better yet intuitive controls for operation. Apple has proven this repeatedly, including most recently the wildly successful iPod. Designers need to keep in mind this broader perspective on what successful product creation means.

Interactive Internet feedback. The Internet has introduced a completely new channel for customer feedback. Intel's® infamous floating-point calculation flaw in its Pentium® processor in 1994 presaged the rising power of the Internet-connected community of customers. Today every company, not just those that serve a community of computer users, can expect Internet feedback.

Engadget, a web magazine that claims an "obsessive daily coverage of everything new in gadgets and consumer electronics" offered a brief review of Unicon System's MBridge™, a Linux based personal media player the company touted as "the key to a complete computer-free life." In less than four hours, nine bloggers posted their own opinions, some dismissing the marketing message and others exclaiming disbelief at the low price. One even offered a link to a personal blog with video footage of an interview with a Unicon® development engineer at the 2006 Macworld expo held in San Francisco earlier in the month. http://www.engadget.com/ 2006/01/31/unicon-systems-mbridge-probably-not-the-key-to-a-computer-fre/#comments.

On the less-focused "J-Walk" blog, a computer book author from Tucson, Arizona, shared his opinions about black and white M&M'S® (see Mars, Chapter 9) and triggered responses from other consumers. "Blog," a word that did not exist in the vernacular a decade ago, now generates 158 million hits in a "Google™ search." Companies that ignore such feedback do so at their own peril.

Responding to the growing influence of the Internet, leading companies actively engage consumers by establishing online customer panels that offer insight through their answers to directed questions from the sponsoring company. Other companies have migrated traditional focus groups to the online world, which lowers the cost of physically assembling a group of individuals. This approach also broadens the companies' reach to groups often unavailable in a geographically constrained model. The times are not only changing, they change daily.

Techniques Requiring a Concept Example

* * *

All of the techniques for market sensing described thus far can be applied without a specific concept in mind. Generally they serve as input to eventual formal efforts at product idea generation, but companies can use them to test a specific product or concept as well. Other techniques, however, work only with a specific, predefined concept. The customer needs to experience and react to the concept for the developers to glean the needed market sensing insight.

"Concept cars" and tradeshows. The most popular exhibits at automotive shows from Tokyo to Detroit to Stuttgart display innovative concepts cars that demonstrate leading-edge automobile design thinking. Although most concept cars never reach production, automotive companies use them to stimulate consumer interest and generate feedback on trends. By going beyond conventional design principles, concept cars also help shape consumer expectations about future design trends and prepare consumers for new paradigms.

At times, the buzz around a new concept car becomes so loud that it can trigger a reprioritization of development plans. The FJ Cruiser from Toyota, presented as a concept car at the 2003 Detroit Auto Show, created so much stir that Toyota inserted a production version into its formal lineup a mere two years later.[7] In fact, tangible concepts in the form of prototypes are the foundation of

---

[7] Volvo C30, Toyota FJ Cruiser headed to U.S.; Mercedes to sell diesel M-Class," *Automotive News*, January 11, 2005.

Toyota's rapid development process. Fast iteration of multiple alternatives allows parts suppliers and Toyota to quickly settle on an optimal overall design for major subassemblies.

The automotive industry does not have a monopoly on the introduction of new product concepts at industry trade shows. The 2500 exhibitors at the 2006 International Consumer Electronics Show in Las Vegas introduced new products and emerging concepts while senior executives from companies such as Microsoft®, Sony®, Dell®, Kodak®, Intel®, and Yahoo®! discussed key industry trends.[8]

Even less well-known industries use the automotive industry's "concept cars" approach and tradeshows as lenses on the future. The International Music Products Association has held annual shows for more than 100 years with recent attendance approaching 80,000[9] while the even older Food Processing Machinery & Supplies Association, founded in 1885, draws around 75,000 industry professionals to see the "latest innovations and technology in the food processing sector."[10]

Prototyping. Further along in the development process, prototypes offer a way to make a product tangible and to test its utility at the same time. A focus on speed to market in the development of new ideas and the refinement of existing ideas, has led to the notion of "rapid prototyping." Although valuable for gaining important customer insight, because prototyping can be expensive, it must be well-managed by development groups. Senior management can add the value of their experience and insight

---

[8] http://www.cesweb.org.
[9] http://www.namm.com.
[10] http://www.foodprocessingmachinery.com.

into the cost-benefit decision between gaining further insight to avoid a market flop and investing the cost and time associated with prototyping.

Mars uses prototypes as part of its consumer testing research. The company's consumer insight experts, a multifunctional group that resides within the research and development organization, has always used traditional focus groups and ethnographic techniques. Recently the group has expanded its toolbox to include prototype and design capabilities. Now it rapidly creates trial products, complete with packaging, for use in consumer tests or to demonstrate a concept to senior management.

Usability laboratories. Usability labs provide another way to observe a product in use. Whirlpool displays its prototypes and current products in space designed to emulate natural settings— the kitchen, laundry room, patio, or garage, for example. In such settings, home economics professionals, designers, and consumers test products through continual use. These lab tests provide insights into a product's ease of use and functional performance and can highlight product shortcomings and lead to potential product enhancements. Whirlpool has numerous usability labs not only in its research and design centers in the United States, Italy, and Brazil, but also in the field, far removed from central R&D. The "Insperience™ Studio" in Atlanta is a prime example. There major customers and suppliers can see new developments up close and personal. Such usability labs, however, do not replace product life-cycle testing and product quality control labs. These testing laboratories assure reliability and endurance and are appropriately distinct from usability labs.

Product use simulations. Product creation teams can also seek to understand customer needs by making concepts tangible through product simulations that demonstrate virtual rather than physical functionality. Used widely in manufacturing industries, product simulations test the usability of practically any product feature from appliance control panels to automobile interior features to farm equipment mechanics and, of course, to aircraft cockpits. Software companies also use simulations to design and market their products. For example, virtual simulations can provide customers a sense of the input and output interfaces before the back-end transaction processor is developed.

The U.S. military has a long tradition of using simulation. Sophisticated tools allow aerospace producers to test the performance and benefits of the proposed product in various user scenarios. Now mimicked in the enormous and sophisticated modern video games industry, complex interactive simulations capture pilot input on the performance of new concepts without risking lives. Other simulation applications model the current strategic and tactical military doctrine of a modern battlefield to test the efficacy of new equipment and tactics. Because the Department of Defense lacks a fully developed in-house capability for such simulation, this tool proves useful both in developing products and in marketing to potential customers.

Boeing and Lockheed Martin have organizational units that use simulation to define and then market their products to Defense Department specialists. For example, Boeing has centers staffed in part by former military leaders who bring with them both credibility and understanding of current military doctrine and future strategic direction. These centers serve all Department of Defense service and agency acquisition groups and NASA for space applications.

Monitor Market Influences

* ** *

Although companies must put the customer front and center in the development of new product ideas, they ignore the influence of other market forces at their peril. Companies should systematically benchmark the offering of competitors.

In the late 1980s, General Motors established the Mona Lisa Center, which displayed cars from Ford, Jaguar, Toyota, and other global category leaders. Unlike the namesake painting at the Louvre, the featured items did not rest behind protective glass viewable only from a distance. Instead, the automobiles sat completely disassembled or literally cut in half (including a $60,000 Jaguar!) for the most intimate view possible. Product creation teams from all over General Motors came to the facility to conduct competitive assessments and seek new ideas.

Customers make buying decisions based on value, a combination of features and price. Competitive assessments, therefore, must go beyond feature-to-feature mapping to include the underlying cost and product architecture strategy. In the late 1980s, a product team developing braking systems for GM used the Mona Lisa Center for a competitive assessment, dutifully measuring and weighing each part. They failed, however, to notice that Toyota used the exact same brake on the Camry and the Celica, offering significant scale economies. Similarly a manufacturer of room air conditioners was surprised when an Asian manufacturer offered to sell them a functionally equivalent low-end unit at a price less than the company's materials cost—excluding all assembly labor and plant overhead—despite the fact that the company regularly conducted competitive assessments and product teardowns of the Asian manufacturer's offerings.

Benchmarking should not be limited to existing competitors. Clayton Christensen has shown that disruptive innovations often come from outside of the normally defined industry boundaries. Initially viewed as inferior tradeoffs by current competitors and their core customers, disruptive technologies eventually come to dominate whole industries and force the complete shuttering of incumbent players. Christensen has documented not only recent instances of this phenomenon in the high-tech computer disc drive industry, but also decades old examples such as when hydraulic earthmoving equipment manufacturers like Caterpillar® completely displaced cable-driven steam shovel producers.[11]

Industry disruptions and influences can also come from other market participants such as channel partners or providers of complementary products and services. For example, Circuit City™ made a strategic decision to drop white goods from its offering, creating a major discontinuity in the appliance industry. Large retailers like Wal-Mart™, Lowe's®, Sears, and Carrefour exert great influence on suppliers and often dictate special functions and features for products in their stores. They even create their own brands. The strategies of both Microsoft and Intel continue to shape the entire personal computer industry. As such companies cannot limit benchmarking activities to competitors alone, but must understand all of the forces shaping the marketplace. The very best benchmarking efforts offer insight into strategies

---

[11] Christensen, Clayton, *The Innovator's Dilemma: When New Technologies Cause Great Firms to Fail*, Harvard Business School Press, June 1997.

and enable companies to predict future competitive actions *before* they are announced. Anticipating and planning for such discontinuities create a potential for competitive advantage.

Listen Well. Act Early

* ✳ *

Sensing customer needs and market opportunities provides the raw material for new product and service ideas. It can apply in the fuzzy front-end for generating new product ideas, but can prove equally valuable in focusing efforts in a more narrowly defined product creation project and in identifying and prioritizing opportunities for minor upgrades to the product portfolio. No magic formula dictates which lenses work best for a given product or in a given context. Through continued use and experimentation, however, organizations will find those that work best for them in their own settings.

The more senses a company uses to tap into customer wants, needs, and preferences, the better. Listening, observing, mapping, analyzing, and most of all creative thinking help identify customer needs and open the mind to new possibilities. Repeated market sensing and thinking yields great new product ideas that can meet the commercial goal articulated by Thomas Edison a century ago, when he proclaimed "Anything that won't sell, I don't want to invent. Its sale is proof of utility, and utility is success."[12]

---

[12] http://www.brainyquote.com/quotes/quotes/t/thomasaed149038.html.

Although inherently "fuzzy," the front-end idea generation process must not be allowed to languish. Companies should always remain vigilant about harvesting concepts at the right time. For most good product concepts, a window of opportunity exists that if missed can mean the difference between success and failure. Concepts need not always be fully ripe before they are harvested. The software industry continuously releases new products and new versions. Most customers now understand that those new versions are probably already in development. In fact many companies including the Pentagon now adopt a spiral development strategy to assure that products are introduced as early as possible and that preplanned improvements are scheduled for future versions. Keys to a successful early harvest include separating invention from execution, providing a clear product architecture, continually renewing the product portfolio, and instilling processes that encourage discipline without crushing creativity. The following chapters expand on how to convert ideas into a competitive product portfolio.

Chapter 3

TECHNOLOGY CAPABILITIES

———— ✳ ✳ ✳ ————

Separating Invention and Execution

Thomas Edison, a man often cited in this book for good reason, highlighted the research laboratory as his greatest invention. His Menlo Park, New Jersey laboratory founded in 1876 began with 20 "earnest men" searching for a solution to what Edison described as "the electric light problem." Edison promised his investors one new invention every ten days and a major breakthrough every six months.[1] During its short but prodigious history, this earliest example of an industrial research facility became known as the invention factory for the wealth of innovations it produced, ranging from the phonograph to the electric light bulb to the vacuum tube.

---

[1] Presentation by Andrew Hargadon, author of "How Breakthroughs Happen: The surprising truth about how companies innovate," (2003), Harvard Business School Press, Boston.

In the tradition of Edison, effective innovators continuously develop technology capabilities that complement the ongoing process of customer understanding and observation. Even though customer needs ultimately define the success or failure of a new product or service, direct observation alone cannot fully define the opportunity space: most customers simply do not know what might be technically possible.

Exploring technical feasibility, however, does not necessarily mean invention from scratch. Many of Edison's Menlo Park inventions reflected creative application of technologies from one industry to another. The phonograph leveraged existing technologies from the telegraph industry. Even with prudent use of existing technologies, the task of developing new technology capabilities proves far less predictable than the disciplined process of new product creation can usually accommodate. Trying to develop a new technological capability as part of a product creation project has destroyed many innovation efforts.

Furthermore, many technical capabilities apply across several business units and require a critical mass to achieve substantive results. Some leading companies, Thomas Edison's own GE for example, continue to invest in a central research organization to build technology capabilities in support of product creation across multiple business units. Others assign responsibility to more decentralized, division-level research organizations.

Research—Not Development

* * *

Although Thomas Edison may have claimed credit for inventing the research laboratory, his original model has evolved dramatically since that time. Wave after wave of technology companies have invested in research laboratories to push forward the science that supports their products, and in the process have continually shifted the scope of the research organization.

Bell Labs, formed in 1925 by American Telephone & Telegraph Company played a compelling role in the transformation of communications and electronics. The foundational mathematics of information theory, the invention of the transistor, the laser, the communications satellite, and the UNIX® operating system all came from Bell Labs. Eleven different Bell Labs scientists won Nobel Prizes for work ranging from the wave nature of matter to radio astronomy to quantum physics. Now a part of Lucent Technologies, Bell Labs employs 9000 people worldwide, including four research and development facilities in China to supplement its core operations in Murray Hill, New Jersey.

Riding a new technology wave, IBM launched its first research laboratory in 1945. Established in collaboration with Columbia University, the Watson Scientific Computing Laboratory began in a renovated fraternity house in Manhattan. In the 1950s, IBM established another facility near San Jose, California, in what has become known as Silicon Valley. In the early 1960s, the Watson Research Center moved to a new location in Yorktown, New York, and later expanded to Hawthorne, New York, and Cambridge, Massachusetts. Like Bell Labs, IBM's version of the research lab continued to support fundamental scientific research. Over the years, IBM researchers won Nobel Prizes for the scanning tunneling electronic microscope and for the discovery of high temperature superconductive materials. IBM researchers also explore leading-edge computing applications ranging from voice recognition technology to copper chips that have more immediate product application potential. Today the sun never sets on IBM research with 3600 staff in Massachusetts, New York, Texas, California, Japan, China, India, Israel, and Switzerland.[2]

A later technology wave led to the creation of the Xerox® Palo Alto Research Center (PARC) formed in 1970. PARC demonstrates the ongoing evolution of the research laboratory ironically returning to its roots of applied research. Although PARC cannot claim any Nobel Prize winners, it did invent such practical devices as the Ethernet and the computer mouse.[3] Since its founding in 1970, PARC has spun off more than a dozen new companies including, ParcPlace which commercialized the object-oriented programming language Smalltalk.org®, Synoptics Communications for fiber optic network media, dpiX® for selling high resolution active matrix liquid crystal display flat panel monitors, and Gyricon Media to commercialize electronic reusable paper. PARC focuses on commercially-driven invention, but stays out of the commercialization process, exemplifying the appropriate role of research laboratories in the modern corporation.

Research organizations separate invention from execution. In today's world, the complex art of invention cannot operate on

---

[2] www.research.ibm.com.
[3] www.parc.com.

a rigidly fixed timeline. Commercial product creation must. The chief technology officer carefully manages the portfolio of technology capabilities to ensure the company maintains an appropriately commercial focus and that no technology enters the product creation process ahead of its time.

Creating the Technology Portfolio

* ❋ *

To maintain an appropriate focus at the research lab, successful senior research managers create a technology portfolio that balances the potential for breakthrough products with the expressed needs of the market. They organize the technology port-folio around a relatively stable set of strategic priorities but with an evolving set of objectives and varying levels of investment. For example, Bruce Vaughn, head of research and development at Disney, selected a set of themes—such as living characters, emersion environments, and special effects—that would remain relevant over many years. The specific projects and objectives evolve over time as new technologies and priorities emerge. A more traditional model organizes around developing technologies like materials, electronics, biotechnology, and nanotechnology. Either approach can work as long as the organizing model allows for the right balance of continuity and flexibility and supports cross-pollination of ideas where appropriate.

The allocation of resources to each area should vary with the level of technology maturity. Rapidly changing technologies demand higher investment levels to respond to shifting product capabilities and market needs. This inevitably means short product life cycles. As the penetration rate of the product or service increases, product technologies mature and standardize, leading to longer product life cycles. Even with mature technologies, smart companies are vigilant for disruptive, technical discontinuities, for example the shift from bias tires to radial tires in the automotive industry or mechanical calculators to electronic ones in business machines.

Making an investment to drive a disruption may not be feasible, because technology disruptions often come from outside of an industry. A company could, however, "take options" on promising technologies by making relatively minor investments. For example, a company could contract at a relatively low cost with an external research center or a collection of individual researchers to monitor new technologies in case they become disruptive.

Even if a company limits its investment in potentially disruptive technologies, part of the advanced technology capabilities portfolio should consist of relatively risky projects that can support step function changes in the existing product portfolio. Other projects should have a much higher chance of success and a relatively low risk of failure even though there may be significant uncertainty about their cost and timing. These are the new capabilities that the organization can count on to be available for new product features on a more predictable timeline. On balance, a research organization can expect a failure rate across the portfolio of around 20 percent for mature products and up to

50 percent for leading edge concepts. Too low a failure rate suggests inadequate investment in breakthrough projects. An inordinately high failure rate suggests the opposite—too many high-risk projects that do not produce returns for the corporation. Remember high-risk projects that offer the potential of high returns are worth the gamble. One or two really big-payoff projects are also generally great morale boosters in the research organization.

Most projects do not explicitly fail: researchers inevitably believe that they can succeed if given enough time and money. Experienced senior management will call the question and force the team to walk away when the amount of investment and the expected duration exceed acceptable standards. Forcing the failure decision without discouraging the researchers requires setting expectations about failure rates upfront. No one wants to have a project cancelled, but understanding the odds of failure helps soften the blow. Also, teams should understand that failures have value because they prove the infeasibility or impracticality of an idea, which in turn can inform future product creation projects.

In the end, managing the investment levels of the technology portfolio requires managerial judgment. Sometimes an opportunity offers such a large potential that it warrants aggressive funding at the expense of other areas. In other cases, management will choose to emphasize some potentially promising areas for a few months to harvest some insight but then move on to new areas based on experienced judgment. Good judgment comes from a flexible mindset and always incorporates into the continued investment decision the latest information from the market and the technology group.

Defining Research Priorities and Projects

* ✳ *

Most strategic research priorities should derive from the voice of the customer research and tie explicitly to the future product portfolio. Some ideas, however, will derive from totally new concepts; things customers would not know are possible. Managing the portfolio of technology projects demands a systematic process that converts the evolving objectives of strategic research into a portfolio of research projects. These projects, in turn, should involve validating inventions on advanced technology demonstrators (ATDs) before introducing the new technology capability into a product creation project. Figure 3.1 shows how the objectives of the hypothetical Omega Corporation, a producer of electronic entertainment systems, not only translate into technology projects, but also their planned linkage to ATDs.

As shown in the figure, a variety of different research disciplines may work in combination to validate a concept. In the Omega Corporation example, electronics, materials science, and bio-tech researchers combine to pursue a project proving a capability in bio-controls.

Since invention does not happen on a fixed timeline, the chief technology manager should review the linkages between objectives and projects at least every six months to ensure that the market opportunity remains. More frequent reviews may be warranted to accommodate a rapid pace of technology change. Less frequent reviews offer little benefit even in less dynamic environments. Given the inherent uncertainty, investment levels need to be

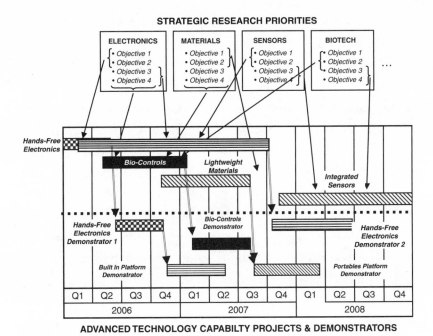

Figure 3.1 Defining Research Projects to Create a Technology Portfolio

assessed frequently to ensure that the investment is justifiable. In less dynamic markets, course corrections may occur less frequently, but management judgment still plays a major role.

Without some forcing function to bring focus, research projects can extend indefinitely. advanced technology demonstrators (ATDs) can fill this need by forcing researchers to validate new technologies on actual prototypes. Although prototypes typically represent a new or existing product, ATDs offer a way to validate new technological capabilities independent of product creation projects. As shown in Figure 3.1, various research projects can feed a specific ATD, which can be developed at a fixed rhythm.

This means that the latest developments can be verified in the context most likely to use the new technology.

ATDs convert an idea into a demonstrated capability by proving the idea on a prototype product "host." The host allows validation within a context that simulates, to the extent possible, the real use environments for the innovation. Although ATDs logically link to specific product creation projects, their development must remain separate. Companies have lost millions of dollars trying to invent during product development. ATDs offer a flexible way to link research projects and product creation projects without requiring that both adhere to the same timeline. Product creation projects should only incorporate technologies that have been proven on an ATD.

Not all new ideas need to be cycled into the research labs and validated on ATDs, which prove most useful for breakthrough or complex concepts. The introduction or substitution of a new component needs to be tested in all simulated use environments, like an ATD, but this can be done through prototypes developed in the new product creation projects. A complex system change— even if the individual components do not represent breakthrough technologies—will warrant the involvement of the research group and the use of an ATD.

The use of an advanced technology demonstrator may appear to slow down the innovation process. Experience has shown the opposite. The research group can take a less mature idea and refine it in the supportive environment of the laboratory without the timing pressures imposed upon a specific product development project. The ATD forces the research group to make the concept real by demonstrating a functioning prototype. The demonstrator also becomes a vehicle to show the organization—including senior management and the customer—the possibilities of the concept. Showing is far more powerful than saying.

Although research efforts have longer time horizons and tend to be more unpredictable than product creation projects, ATDs should include some level of schedule discipline. Producing ATDs on a fixed schedule simply requires accepting that the actual content of a specific ATD might not incorporate all of the advanced technology originally envisioned. A subsequent demonstrator, also developed on a disciplined schedule, would then host the delayed technology capability. Some capabilities still under development can be realigned for future ATDs. The clear linkages—but flexible timing—between technology objectives and the research program, coupled with the use of ATDs to validate concepts, instills a realistic level of discipline in the unpredictable world of invention.

Drawing Technology Roadmaps

* * *

Technology demonstrators alone do not produce returns to the company. New technology capabilities only create value when they come to the marketplace through the product portfolio. Technology roadmaps explicitly link the research function, the product creation teams, and the product in the market. Figure 3.2 offers a hypothetical example: the Omega Corporation's development of hands-free, voice-activated controls for use in its line of entertainment systems.

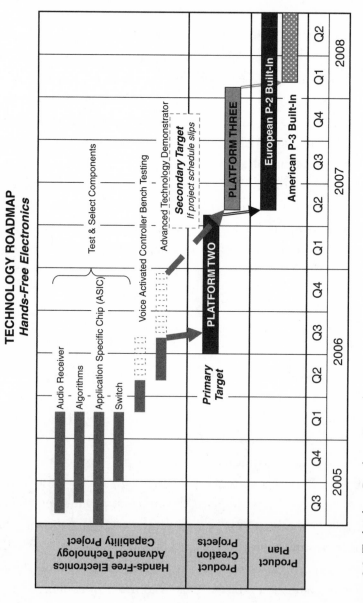

Figure 3.2 Technology Roadmap Example

Just as the roadmap links the technology project to the product portfolio through the product creation projects, it also establishes—in advance—a back-up or contingency plan if the invention effort takes longer than expected. In the Omega Corporation example, a European built-in product serves as the primary target for the voice-activated, hands-free controller. If, however, the technology demonstration slips, the new technology can be incorporated into a product creation project expected to launch less than a year later in the American market. Chapter 5 explains in more detail how to forge these links and ensure that the technology-forward view meshes with the market-back view.

Forming Research Alliances

* * *

The portfolio of advanced technology capability projects need not be completely in-house. Xerox spun off PARC in 2002 to provide research service not only to Xerox, but to other companies as well. This better leverages the technological depth of PARC and exposes the researchers to more external stimuli, which can lead to even greater breakthrough thinking. Many companies turn to external research organizations, such as government sponsored organizations, private consortia, and universities. Although accessible to any business, external research resources come at a cost, both direct and indirect.

The U.S. government operates many research entities that could be of use to commercial businesses. The government encourages these entities to identify and support "dual use" technology for commercial applications where possible. The Air Force Research Laboratory at Wright-Patterson Air Force base in Dayton, Ohio, and the Naval Research Laboratory in Washington, D.C., study materials, advanced manufacturing techniques, and many leading edge technologies. Although commercial enterprises may have an interest in such research, getting access to it requires determining the capabilities and forming working relationships with key researchers to facilitate technology transfer. The 36 federally funded research and development centers (FFRDCs) at universities and corporations sponsored by nine U.S. government entities offer another potential source for external research and development expertise. These organizations do much of the federally funded applied research, but also offer the private sector opportunities to collaborate in some areas.

For example, the Department of Defense sponsored Software Engineering Institute at Carnegie Mellon University in Pittsburgh, Pennsylvania, researches tools and best practices for developing and using software, obviously of potential value to all industries. The Fermi National Accelerometer Laboratory conducts leading edge research into fundamental physics and the National Renewable Energy Laboratory explores capturing energy from sources like solar, wind, geothermal, biomass, and fuel cells.

Private research corporations manage many of the FFRDCs, which may make them even more accessible to corporate researchers. For example, Mitre Corporation, a not-for-profit organization headquartered in Bedford, Massachusetts, and McLean, Virginia, operates three FFRDCs. It also conducts its own research and licenses many of its technological innovations. Battelle of Columbus, Ohio, operates five FFRDCs, but also has commercial operations including a $150 million venture fund for investing in start-ups in five of its targeted fields: life sciences, information technology, homeland security, energy, and advanced materials and nanotechnology.

Other options include private research entities like the previously mentioned PARC or the Hughes Research Laboratory (HRL), originally part of Hughes Corporation but now funded by Boeing, General Motors, and Raytheon. Similar to the joint ownership model of HRL, many industries sponsor research consortia like Electric Power Research Institute (EPRI), funded by the electric power industry; Gas Research Institute (GRI), funded by the gas industry; and Sematech (computer chip fabrication equipment), funded by the semiconductor industry. These consortia conduct research and development in support of their members, yet some research will often have applications outside the sponsors' industry. For example, Maytag® worked with EPRI on a microwave clothes dryer.

Many of these organizations allow private sector researchers access to their laboratories and use of their highly specialized capabilities. Such access is not without cost. To realize effective technology transfer, in terms of both financial and human resources, companies should prioritize the available opportunities and then commit resources to the external entity and manage it in a similar way to their in-house activities. Accordingly, such efforts need clear plans that are no different from the funding and goal setting of internally funded R&D. Without such alignment, the technology transfer will be difficult if not impossible. In addition, such collaboration also requires clear, mutual understanding of shared investment strategies and ownership of proprietary technology.

Often a university houses an industrial applied research institute, for example, the International Motor Vehicle Program at the Massachusetts Institute of Technology or Michigan State's Materials Laboratory and Purdue's Nanoscience Center. In yet another variation on university-based research and development resources, a company can work directly with academic researchers at leading institutions by providing a small grant to the university or offering a consulting contract directly to a professor working in an area of research interest. Universities strongly encourage faculties to acquire grant money and to consult in areas that will build the individual's knowledge base.

Despite the wide array of existing external resources available to corporate researchers, most of them have goals and cultures that differ significantly from a typical commercial business. These outside entities tend to work within longer timeframes, in a more relaxed atmosphere, and with a less urgency. Even when a great sense of urgency arises, it may very well stem from researchers' desires to gain more knowledge in their specialty rather than from meeting the commercial goals of a business "customer."

Even if an organization chooses to do all development work in-house, prudent managers will ensure that the internal research professionals actively explore and monitor outside developments. Too many corporate research organizations become insular and constrained by the "not-invented-here" syndrome. Researchers should seek outside collaboration supported by a modest budget to develop relationships with key organizations and activities to keep in touch with the outside world. Attending professional meetings can help build important links with individuals and organizations doing research that aligns with internal needs.

It also helps internal research professionals build the social connections and intellectual capital that contribute to personal and corporate success.

Once an opportunity arises to acquire external technical capabilities, much work remains to make it useful to the business. Whether for-profit or not-for-profit, external research organizations view corporations as potentially deep-pocket providers to fund their efforts. Although providing modest funds to external researchers may keep staff members attuned to the outside world, an external research effort should not be in a position that could block the critical path for near-term commercial plans. External efforts, however, can help identify young talent and stimulate fresh ideas for internal researchers.

Finally, companies should always approach external research efforts with a keen eye on intellectual property issues. Although time spent on a legal review of intellectual property may appear to impede progress, external collaboration demands the involvement of legal counsel to protect proprietary intellectual capital.

Considering Organizational Models

*** *** ***

Regardless of the extent of external research services a company uses, to compete on the basis of new product creation it needs an internal research organization of some sort. At a minimum, a research group explicitly decouples product creation from concept validation. The separation of invention and development serves as a fundamental tenet of effective new product creation. Without such separation, development projects can extend beyond planned time lines, adding cost and missing market opportunities.

Research organizations also provide a critical mass for building technology capabilities that apply across business units and product lines such as biochemistry, physics, electronics, and standard development tools such as environmental test chambers and computer aided design (CAD) software. For example United Technology Corporation, which operates a highly decentralized business model, maintains the United Technology Research Center (UTRC) to lead the corporation in the key areas of basic technology and analytic modeling of complex systems and components. The business units also "contract" with UTRC to support advanced technology development requiring major breakthroughs like the coated steel belt technology deployed in the Otis Gen2™ elevator. Staffed with 450 people out of UTC's global total of 210,000, the research center represents only a fraction of the enterprise's total technology resources. Nonetheless, it remains a vital contributor of innovation processes and technical capabilities.

Large organizations may have research units at the business unit level, but most of these decentralized technology resources typically focus on bringing a product or service to market as well as ongoing product support. As such, decentralized technology staff members tend to be practical and more short-term goal driven because they focus on disciplined engineering rather than technological breakthroughs. For example, most of Disney's Imagineering organization develops new park attractions from clever combinations of existing technology. Imagineering's small research and development group, however, focuses on breakthrough concepts like smokeless fireworks and robotic characters, which can eventually be deployed in the parks, but not on a rigid, fixed schedule.

Geographically dispersed business units derive particular benefits from centralized research management, often leveraging technology to achieve virtual rather than physical integration. UTC's Carrier business operates geographic profit and loss units but seeks to integrate its development efforts across the world by positioning different research facilities as centers of excellence. Many companies have begun expanding their research organizations to low labor-cost countries like India and China. Although these countries undoubtedly possess exceptionally well-trained technical resources, some companies struggle to integrate offshore research into their domestic function because of cultural and communications challenges. Longer term, these low-cost facilities may prove immensely valuable, but in the near term they continue to require active involvement from senior research management to ensure effective integration.

"Skunkworks," a term now used generically, originated as the informal name for the Lockheed Advanced Development Programs Unit, which was chartered in the mid-1940s as a top secret development program to design the first fighter jet in a mere 143 days. Today, the term refers to any development group operating outside of the normal corporate hierarchy and process discipline. Skunkworks offer the potential to create a highly motivated team, unconstrained by bureaucracy and narrowly focused on a single and usually high-pay-off objective without distraction. Although a potentially powerful tool for major breakthrough efforts—especially on a tight timeline—skunkworks can become disconnected from the day-to-day customer and can disrupt the organization at large because of a sense of elitism. Given the pressures of this environment, skunkworks can prove difficult to sustain and should be used sparingly.

Regardless of the research function's organizational structure, collaboration across units is essential. Periodically rotating people among organizational units builds crucial linkages and helps ensure effective communication and respect for the capabilities of other experts in the organization. Rotation among central research and the business units proves particularly important.

Corporate research units tend to focus on generic areas like materials and electronics or emerging technologies like nanotechnology and genetic engineering—capabilities that may require substantial time to yield a commercial return. As a consequence, decentralized business units often view corporate research organizations as frivolous and unnecessary. Conversely, research and development groups at the business unit level often become so closely tied to the product and its day to day issues that they struggle to generate paradigm shifting concepts. Rotating a few respected people among these groups keeps extreme views from dominating, and cross-pollinated staff facilitates integration across organizational units far better than any process.

There is no one-size-fits-all organization model for building technology capabilities. Inevitably, every company faces tension and tradeoffs between the centralization needed to build critical mass and the decentralization needed to ensure ownership. Senior management must decide and ultimately ensure that the processes support the organization model and minimize the negative tradeoffs.

Addressing the Softer Side

* ✳ *

Managing research groups productively presents a real challenge for business leaders. Many do not understand the process or the motivation of the people who do the work. Although difficult, managing researchers can be the most rewarding and enjoyable aspect of innovation management. The people in these organizations tend to be very smart, creative, and motivated by things other than money. Seeing their inventions routinely get to market provides a source of pride. They also enjoy the freedom to pursue discretionary efforts with some fraction of their time. For example, 3M has a policy of allowing its researchers to deploy 15 percent of their time on projects of their own choosing.

Researchers also respond well to recognition ranging from simple thank-you notes from senior executives to the formal acknowledgement of patent awards at professional gatherings. A former Apple development engineer noted that his first suit was a tuxedo. The reason? Daily attire at Apple is decidedly informal, but the company regularly hosted black tie events to celebrate patent grants. And though they value their freedom, researchers also need to stay connected to the business and its customers through business updates from senior managers and occasional field work with product support engineers.

Money may not be the prime motivator for researchers, but they definitely respond better if they have clear career ladders. At United Technologies Research Center, for example, rather than increased managerial responsibility typical in other parts of the business, professional advancement for researchers can include designation as a Corporate Fellow. Despite their unique skills and motivations, research professionals will generally produce beyond expectation when genuinely appreciated.

Although research groups value their independence, management must delicately steer them toward relevant work. This means giving strong direction regarding the role of researchers not only in pursuing breakthrough concepts, but also in translating those concepts for the market. Ultimately the researchers' output must be linked to future product improvements for the marketplace and supported by an investment strategy. At times the researchers should be deployed in support of product creation teams. The right mix of discretionary time, directed research, and product creation efforts should produce a team of researchers that consistently produces market results and feels ownership for the current and future health of the product in the marketplace. If, however, more than 20 to 30 percent of researchers' work supports existing product in the market, the appropriate balance has been lost. At that point, the advanced technology group has become more like a product support group.

Another critical consideration in managing a research group involves capturing "tribal knowledge." Few companies have formal processes for capturing and sharing such tacit knowledge. Fewer still do it well. However, tribal knowledge can prove to be the core competence of a company as it may express itself in the unique "touch and feel" of a product like Apple's vaunted industrial design expertise or the durable design of John Deere tractors. Given tribal knowledge's potential to serve as a barrier to entry for copycat competitors, senior management should explicitly seek to nurture and maintain it. One of the best techniques involves pairing young researchers with older ones who share tribal knowledge in an apprenticeship relationship. Chapter 6 offers even more guidance on how to capture the value of this intellectual asset.

The final critical issue in managing research laboratories arises from the organization's natural tendency to "invent" rather than "re-use." Innovation and technology efforts don't always have to be proprietary breakthroughs. Companies can often find technical solutions in other industries that can be leveraged for their own needs. Craig Wynett, who leads the Future Growth Initiatives for Procter & Gamble, dismissed the traditionally lauded activities of brainstorming and creative thinking in a recent presentation at University of Virginia's Darden Graduate School of Business. Instead he focuses his teams on solving new problems with existing solutions from outside their industry.

The natural tendency of those drawn to work in research laboratories is to invent new solutions. Senior management must constantly challenge researchers to avoid becoming so enamored with technical breakthroughs that they fail to focus on solving problems. In a similar vein, researchers can focus too much on new technical concepts and lose sight of the bigger picture. Managers can counter this tendency by posing a challenge to the technical expert: "Let's assume you will succeed in achieving the desired technical development. Now, tell me why I should care." The discussion that follows such a challenge should then focus on the relevance of the outcome more than the significance or difficulty of the innovation or technology. Projects that have relevance can always find a home. Those that produce only technology for technology's sake rarely do.

Appropriately managing the research organization requires separating invention from execution while keeping technology efforts attuned to market needs. Chapter 4 expands on this theme by describing the design and management of a stage-gate process for product creation that leverages advanced technology efforts without allowing them to put project success at risk. Chapter 5 explains the overall process for integrating the resulting portfolio of product creation projects with the portfolio of advanced technologies to generate a competitive product portfolio.

Chapter 4

PRODUCT CREATION PROCESS DESIGN

———— * * * ————

Leveraging Discipline and Judgment

In product creation, achieving the proper balance between disciplined execution and practical, creative flexibility demands expert judgment and leadership by experienced executives. Former automotive industry executive, Lee Iacocca, President of Ford Motor Company and later Chief Executive Officer and Chairman of Chrysler Corporation, got most of that right most of the time.

The "father" of the Ford Mustang, Iacocca epitomized the kind of "car guy" who built the industry through a deep understanding of and superb intuition about automotive consumers. As the CEO of Chrysler, he helped create a whole new category by marshalling the introduction of the first minivan in 1983 and triggering the revival of convertibles in the 1990s. (Detroit had virtually stopped production of convertibles in the 1980s because of liability concerns.)

71

More than simply an executive visionary, Iacocca played an active leadership role in the product development decisions made during stage-gate[1] reviews. For example, when he saw an early clay model of the Chrysler Neon, introduced in 1995, Iacocca sensed that the headlights just weren't right: the rectangular lamps created a harsh image, inconsistent with the fun, friendly Neon concept. Iacocca sent the team back to the drawing board to develop the soft, rounded headlights that ultimately became a hallmark of one of Chrysler's most successful small cars ever.

Iacocca demonstrated through his leadership that process alone is not enough. Successful product creation demands executives who can bring their experience and intuition to the business, and, at the same time, avoid quashing the creativity—and sense of accountability—of the project team. They provide leadership and guidance, but resist making unilateral decisions or forcing outcomes that strip control from those responsible for fulfilling the mission.

The following discussion crystallizes many of the lessons learned by executives and senior managers who have been in the trenches and know from experience what works and what doesn't in product creation. Specifically, it focuses on the essential characteristics of and requirements for developing an effective stage-gate product creation process. At the same time, it emphasizes the responsibility of senior management to take an active

---

[1] Stage-Gate® is a registered trademark of the Product Development Institute Inc. The term is used here in its generic sense to denote a phased product creation process that includes stages and tollgate reviews.

leadership role, to support an innovation culture and to assure experienced-based judgment throughout. Without effective leadership, even a well-designed product creation process can become a bureaucratic straight-jacket stifling creativity, rather than an entrepreneurial life vest supporting and sustaining strategic product creation.

Designing a Stage-Gate Process

* * *

Many successful corporations use some form of stage-gate project management process that breaks product creation into a series of well-defined phases, typically four to six. Each phase or stage consists of a series of activities to ensure completion of the project as quickly and efficiently as possible. The stage-gate process also includes strategically placed tollgates that force critical thinking and assessment of work done to date before a project can receive the go-ahead to the next stage.

The early phases of such processes focus on creativity—exploring options and considering possibilities—while the later stages focus on the details of design and execution. The names given to the stages vary by company, but usually offer a clear idea of the key priorities of each process phase. For example, an excellent textbook, *Revolutionizing Product Development* by Steven

Wheelwright and Kim Clark of the Harvard Business School describes five phases:[2]

- Idea generation

- Product definition and selection

- Design and build prototypes

- Pilot production

- Manufacturing ramp-up.

Stephen Rosenthal of Boston University also describes a five-phase stage-gate process in his *Effective Product Design and Development: How to Cut Lead Time and Increase Customer Satisfaction*,[3] but assigns slightly different titles and priorities to each phase:

- Idea validation

- Conceptual design

- Specification and design

---

[2] Wheelwright, Steven and Clark, Kim, *Revolutionizing Product Development Quantum Leaps in Speed, Efficiency, and Quality,* Free Press, 1992.

[3] Rosenthal, Steven, *Effective Product Design and Development: How to Cut Lead time and Increase Customer Satisfaction*, McGraw-Hill Professional Publishing, 1992.

- Prototype production and testing

- Manufacturing ramp-up

How Many Stages and Gates?

* * *

The appropriate number of stages for a company's product creation process depends on the industry and the product. A company that produces industrial products and has a small customer base will probably have more directed customer input than a consumer products company with widely diverse customers and product lines. Therefore, the more tightly focused company might truncate some of the up-front product creation stages. For example, a small plastic-injection molding company that develops new product applications for its current customers using existing process technology might not need separate ideation and concept development stages.

In some cases, the customer's internal requirements influence a company's product creation process. For example, the U.S. Department of Defense's acquisition process adds a great deal of complexity—including more tollgates—for the defense and aerospace industry. Similarly, highly complex projects or those with uncertain investments may require more tollgates to manage the increased risk exposure. Projects that have long development lead times or extend across several years may also warrant more tollgates.

While adding more tollgates might seem the solution in these circumstances, formal tollgates inevitably consume valuable time as the team prepares for the review. Positively, such preparation forces the team to pause and reflect on the business proposition and their progress. The questions is, "How much reflection is valuable and at what point does it becomes excess overhead?"

Instead of adding more stages and tollgates, companies should consider the alternative of using intermediate milestones. Tollgates indicate clear "go/no-go" decision points. A milestone, on the other hand, offers the opportunity to ensure that the project stays on track and on schedule, and that resources are allocated appropriately. Milestones should occur at critical junctures only to minimize the turmoil and uncertainty for the project team. Whirlpool, for example, has a capital funding milestone triggered by the successful completion of an important, pre-defined project deliverable. This serves as a vital monitoring point in the design and build phase without the need for a full blown go/no-go tollgate.

A Versatile Model

* * *

The stage-gate product creation model presented in this chapter condenses the process into four phases and uses terminology that can apply to both manufacturing and service industries:

- Ideation

- Concept development

- Design and build

- Launch

Consolidating the number of stages into four reduces the overhead usually associated with passing through tollgates, the checkpoints where a team must present a compelling case to executive management to move the project forward. As such, tollgates force project teams to resolve unanswered questions and to consider tradeoffs. Tollgates also mitigate the natural tendency to avoid or delay decisions in order to leave open numerous options.

Despite their positive impact, tollgates also generate a flurry of lower-value activity as teams develop reports and executive presentations to document their progress and to gain approval for proceeding to the next stage. In some companies the sessions become so elaborate that teams expend disproportionate time, energy, and resources talking about the project rather than executing the work. The four-phase process presented here helps limit such low-value activity.

Ultimately, the nature, scope, and corporate context will determine the optimal stage-gate process choice. But, in no case should a company have different stage-gate product creation processes for different projects. Multiple processes cause confusion and eliminate the benefits that accrue from a common language and process discipline embraced enterprise-wide.

Deconstructing the Model

* * *

Regardless of how many phases a stage-gate process may have, each phase should be separated from the succeeding phase with a well-defined decision point—the tollgate—as shown in Figure 4.1. In the model presented here, the decision point for the ideation phase is called the idea screen. At this tollgate, the project team must demonstrate that an idea has a reasonable chance of resulting in a profitable business opportunity.

Having successfully passed through the idea screen, the project team embarks on the concept development phase to explore multiple options for pursuing the approved idea. Ultimately, the team confronts its second tollgate, concept selection, which narrows the various options under consideration to one (or in some rare cases, a few) to pursue further in the design and build phase.

Design and build is the "engineering" phase of the product creation process. Here the team does the detailed work of defining how to deliver concepts practically in an operating context. Appropriately, the tollgate for design and build is called launch readiness, which ensures that the team has dealt with all of the necessary details to ensure the successful launch of the product or service.

Two decades ago, many companies considered launch readiness the final tollgate. At this point, the development team typically "threw the project over the transom" to operations, which then had to sort through a host of unaddressed problems to create a repeatable manufacturing or service delivery process. Although most companies today recognize the importance of ensuring

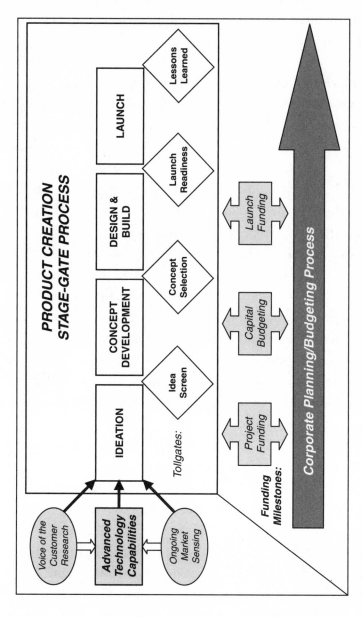

Figure 4.1 Product Creation Stage-Gate Process and Key Linkages

process continuity through launch, rarely do they systematically execute the final tollgate at the end of the launch phase, here "lessons learned".

Critical feeds. Despite the visual simplicity of its structure, the basic four-stage process shown in Figure 4.1 also has many critical interactions with other corporate functions and processes, including marketing, research and development, and finance. Note on the left-hand side of the figure, the importance of two concepts described in Chapters 1 and 2: the ongoing exploration of the voice of the customer and external market intelligence, including competitive benchmarking, all of which can feed directly or indirectly into ideation. Note also that the advanced technology efforts discussed in Chapter 3 also feed ideation in varying degrees. This pre-work to the formal product creation process helps shape the opportunity space from which the specific characteristics, functions, and features of the new product or service will emerge. The most successful projects begin with as large an opportunity space as possible, constrained only by affordability, the company's capability to deliver, and brand-relevance.

To tap into the voice of its customers, Mars conducts ongoing ethnographic research into consumer behavior independent of any particular project. The research findings—such as observations about how children snack during car trips—can serve as indirect input into an advanced technology development effort in packaging or feed directly into a new product development project for M&M'S™. In fact, insight gained from such research led to the development of various types of re-sealable packaging.

In another example, Carrier Corporation, like many companies, regularly scans competitive offerings to identify potential gaps in its product lines, information which again can feed indirectly into technology development efforts or directly into product creation projects depending on the magnitude of the identified gap. A simple line extension typically requires no advanced technology capabilities, but may have a tightly defined cost target based on the competitor's pricing. A major breakthrough, such as an Internet-enabled remote monitoring system, might require advanced technology development efforts before initiating a specific product creation project.

This mix of direct and indirect feeds into the ideation process—often referred to as the "fuzzy front-end"—presents one of the first challenges for the executive. Many ideas ferment in this informal zone for a long time prior to official initiation as a development project. While fermentation is critical in the production of a fine wine or spirit, it can also result in a rotten mess under the wrong conditions. A procedures manual cannot definitively determine the precise point for shifting a project from the fuzzy front-end to the formal product creation process. Such decisions require expert managerial judgment. Experience suggests that given the choice, management should err on the side of an early harvest of creative thinking and quickly impose the discipline of ideation and the idea screening tollgate. This not only avoids needless delays in pursuing good ideas, it also prevents wasted efforts on irrelevant ones.

Functional interfaces. Arrayed along the bottom of Figure 4.1 are critical corporate planning and budgeting processes that must interface with product creation. Too often, discussions of stage-gate processes ignore the reality that these core business processes, by controlling funding levels, often determine the rhythm and timing of product creation projects. A well-designed stage-gate process acknowledges the importance of establishing funding milestones and also provides the flexibility to address funding issues as needed independent of the tollgates or the corporate business process calendar. Similarly, corporate budgeting processes should allow some degree of flexibility to permit product creation to occur at its natural pace.

Considering concurrence. Although not shown explicitly in Figure 4.1, concurrent development presents a separate but related set of challenges. While most product creation activities fit logically and neatly into a single stage, other activities may need to be moved forward or accelerated to hit later deadlines.

For example, the team might address long lead-time activities like tooling or advertising in the concept development phase even though the tasks themselves would logically fall into design and build if lead-time were not an issue. In other cases, initiating an activity ahead of time could prove more efficient. For example, although prototyping activities normally fall in the concept development phase, an initial, inexpensive prototype created in the ideation phase might generate additional, highly useful consumer insight prior to the idea screen tollgate.

Flexibility and risk mitigation. While a rigid stage-gate process that inhibits flexibility and practical tradeoffs might deserve criticism, such a process design does not condemn the stage-gate concept in general. Wise practitioners understand that the stage-gate process provides a useful guide, not a definitive algorithm constrained by the laws of physics. Here once more, executive leadership adds tremendous value by identifying the need for—and facilitating the approval of—practical exceptions to the standard product creation process.

In addition to allowing flexibility to respond to the unexpected, the stage-gate process should also be designed to identify and mitigate risk. At each stage of the process, contingency plans should be in place to address any risks that could jeopardize the successful completion of the project. For example, as discussed in Chapter 3, a product creation project might anticipate using a new technology capability from an advanced technology demonstrator. But, given the uncertainty of such advanced technology efforts, the project plan should have a proven technology option as a backup plan. The need to mitigate the risk of linkages between advanced technology capability projects and product creation projects presents the most obvious example of the need for contingency planning. All product creation projects face uncertainty—from technical glitches to tooling delays to supplier failures to marketing problems—and accordingly need contingent "work arounds" identified upfront. Experience proves that a project plan that assumes that all will go according to plan reflects poor thinking by a naïve project team.

In another context, flexibility gets to the true heart of developing an effective product creation process for a specific company. Starting with the model presented here, or another generic stage-gate model, individual companies should define the product creation activities and deliverables appropriate to their specific business. To help in adapting the four-stage model, the following sections offer more detailed discussions of each of the four stages and associated tollgates.

Ideation

* * *

IDEATION	CONCEPT DEVELOPMENT	DESIGN & BUILD	LAUNCH

Ideally, the ideation phase starts with a compelling business objective or vision articulated by senior management. As discussed in Chapter 2, the task of crafting the vision might initially fall on an innovation cell before a product creation project team is formed. For example, the Pedigree® pet food division of Mars initiated an effort with a broadly conceived goal of "transforming the dog feeding experience." Such a vision creates a vast opportunity space in which the team can consider any number of concepts. In this instance, the ideas could range from creating entirely new food products to developing innovative feeding devices. In each phase the team and the entire organization must understand clearly the purpose of the stage and the expected deliverables.

To help guide this type of thinking, figures shown in this chapter identify typical problems that might be encountered. The first, Figure 4.2, relates to the ideation phase. While the examples are more illustrative than prescriptive, they give a clear indication of the types of deliverables that have proven valuable, and in some cases essential, in practice.

For the ideation phase, where the purpose is to identify a product idea that will result in a good business proposition, a few basics are essential: understanding customer need, market assessment, and

PURPOSE OF IDEATION STAGE	IDEATION STAGE DELIVERABLES
Develop a product idea that has a reasonable chance of resulting in a good business investment	• *Demonstrate a compelling customer need* • *Evaluate market potential* • *Develop rough business model* • *Identify concept development resource requirements*
TYPICAL IDEA SCREEN TOLLGATE ISSUES	
• *Insufficient market potential* • *Potential return does not justify investing resources to flush out* • *Idea not supported by the corporate or brand image* • *Competitor's advantage too difficult to overcome*	

Figure 4.2 Ideation Stage

a preliminary assessment of the potential business return. By the same token, Figure 4.2 also identifies issues that may prove problematic, if not insurmountable.

In ideation, for each fruitful idea, the team must examine the competitive environment, estimate the size of the opportunity, and identify the critical challenges and risks it entails. Typically this phase generates sketches, mock-ups, or virtual prototypes of two or three product concepts in the targeted opportunity space. Ultimately, the team should enter the idea screening tollgate with a preliminary business plan covering the market opportunity and the resources required to move forward. While Figure 4.3 graphically shows the kind of content that the ideation phase business plans should cover, it should be viewed as representative rather than as a comprehensive list.

Market Assessment	Financial Projections	Risks and Contingencies
• Description of customer need • Description of target customer • Description of key product attributes • Market size • Competitor positions • Marketing challenges	• Target price points • Cost targets • Required capital • Projected earnings potential and return on investment	• Market risks and concerns • Potential technical issues • Financial implications • Recommended contingency plans

Figure 4.3 Ideation Stage Business Plan Content

The ideation phase should consider a broad range of options and characterize each product idea in enough detail to assure the team and management that it offers a good business proposition. These descriptions, however, should not attempt to tie down the details of the product or service, but simply define ranges of key metrics and key challenges to be addressed in future stages. For example, in developing the Prius, Toyota set targets for fuel economy that stretched the team but allowed it to initially explore both hybrid and conventional engine technology options.[4]

At completion of ideation, the idea screen tollgate provides the opportunity for management to approve the business proposition and ideas to be fleshed out in the next phase, concept development. Senior management plays a critical role here in challenging the team to focus on the business economics, not just a "cool" technology or a "great" consumer value proposition.

---

[4] Liker, Jeffrey K., *The Toyota Way: 14 Principles from the World's Greatest Manufacturer*, McGraw-Hill, 2004.

Such attributes may offer an interesting starting point for fleshing out ideas, but do not guarantee a sufficient return. For example, the Internet boom and bust at the turn of the millennium offers clear evidence that technology and consumer appeal, while intriguing, may not suffice on their own to deliver favorable financial returns.

The experience of Webvan, the online grocer, serves as a cautionary tale for any executive or project creation team tempted by technology and consumer appeal. Webvan managed to spend $500 million dollars before going bankrupt despite a loyal (though unfortunately small) base of customers who adored the company's technological solution. In fairness, Louis Borders, cofounder of Borders Books® and the originator of the Webvan business concept recognized that the business plan represented huge risks. When a venture capitalist excitedly asserted that "this business could be worth $1 billion," Borders corrected him by explaining, "No, it's going to be worth $10 billion...or zero."[5] Appealing as such large upside gains appear, few companies can afford to make such bets on unproven concepts.

To avoid, or at least bring into clear light, similar pitfalls, senior managers should draw on their experience and wisdom to assess the kind of risk inherent in any innovative concept. Though rarely will they face a bet-the-company strategy, the prudent approval of an occasional high-risk bet can lead to breakthroughs that dramatically reenergize a business. Figure 4.4 offers a set of typical questions that senior managers can pose at the idea screen tollgate.

---

[5] Randall Stross, "Only a bold gamble can save Webvan now," Wall Street Journal February 2, 2001.

What are the compelling customer needs and benefits?
- What current products address these needs?
- What are competitors' strengths with their products and capability?
- What new or different benefits could our product provide?
- Does this product fit our brand/Corporate Image?
- Is this concept part of an external mega-trend?
- Does the concept have strategic value?

Can we have a competitive advantage?
- How can we get to market quickly?
- Is there potential for sustainable competitive advantage?
- What are the strengths and weaknesses of current and potential competitors?
- Can we have a channel or trade advantage?
- Is there a purchased product or licensing opportunity?

What is the idea's potential business benefit?
- How big could the market become?
- What is the expected new revenue growth curve?
- Is the required infrastructure to support the market in place? For example, are we creating CDs without an installed CD player base?
- What are potential margins?
- What major investments are required?
- Is the opportunity financially attractive?
- How does this project fit with our product portfolio strategy?

Do we understand the risks?
- What resources are required for the concept development phase?
- What early experiments could enhance our understanding?
- Is this an "all or nothing" bet?
- Do we have the right competencies to address this opportunity?
- Do we have the technology in house?
- Does our current supply base capability support this concept?

On balance should this project proceed to concept development?

Figure 4.4 Idea Screen Tollgate Questions

Concept Development

* ✳ *

IDEATION	CONCEPT DEVELOPMENT	DESIGN & BUILD	LAUNCH

The purpose of the concept development phase is to develop the product concept in enough detail that it satisfies a compelling customer need, has a viable market, meets brand image, is saleable, and represents a good business proposition. Broadly speaking, the concept development phase should resolve the "knowable" risks in the project and its feasibility overall: Does the customer want it? Is the technology available? Do we have the capability? Can we hit cost targets? Figure 4.5 lays out the purpose, typical deliverables, and potential problems encountered in concept development.

Because it requires answering tough questions, concept development demands the highest degree of tough-minded executive leadership. In this phase, the product creation team takes a rather vague opportunity from ideation and determines all the necessary detail to flesh it out with as much certainty as possible. This is the time to push for greater confidence, not faster resolution of issues. Predicting the speed at which questions of concept feasibility can be resolved—and by extension the duration of this stage—typically produces the greatest variance in total elapsed time of any of the stages.

This phase should also generate the greatest number of tollgate failures: the cost of going to the next stage increases dramatically with the commitment of major capital and the expansion of project

Purpose of Concept Development Stage	Concept Development Stage Deliverables
Develop and then screen potential concepts to capture the identified market opportunity	• Validate customer benefit • Complete market assessment • Verify the capability to produce or acquire • Test concept prototypes in all use environments • Finalize preliminary business case • Identify design and build resource requirements

Typical Concept Selection Tollgate Issues
• Insufficient projected return on investment • Beyond the company's core competencies • Too much technical risk for development • Major change in market or competition

Figure 4.5 Concept Development Stage

team resources. Senior management must probe deeply to ensure that knowable risks are not assumed away in the rush to pass through the concept selection tollgate. When necessary, management must have the will to stop or redirect a concept at this point despite the sunk costs to date, unfortunately not a natural tendency for managers anxious to pursue continuous innovation.

Consider for example the Crusader, a development program initiated by the U.S. Army in 1995 and terminated in 2002 after an expenditure of $2.8 billion. Envisioned in the concept phase as a mobile, rapid-fire howitzer capable of delivering four-inch ordinances over a distance of 30 miles with accuracy within 100 yards of the target, the Crusader concept focused on maneuverability and speed rather than precision. Munitions designed for the Crusader could destroy a city block and the vehicle would require far less manpower than weapons capable of a comparable rate of sustained fire.

By the time the first prototype successfully demonstrated the rapid-fire technology in 2002, military needs had changed dramatically. Originally designed for high intensity European warfare against equally sophisticated Soviet forces of the cold war, the Crusader did not fit the challenges facing the military following the attacks on the United States on September 11, 2001. In the war on terrorism, ground combat takes place with a less sophisticated and concentrated enemy at close range. Deciding to terminate the Crusader before moving to the next phase was painful, but as Secretary of State Donald Rumsfeld testified before the U.S. Congress, it freed up several billion dollars in resources and enabled an accelerated shift to more precise weapon systems with even greater maneuverability[6]. For this program, the market had changed.

During concept development, senior management must challenge project teams to ensure that the technology has been adequately proven. This is critical to ensuring that the next phase, design and build, can focus on engineering execution rather than technology development. Retrospective analysis of failed projects often tracks back to concept development. Ask experienced practitioners about their worst development failure, and most will describe a project that passed into design and build with an unproven technology or a poorly defined concept. Prototypes tested in the use environment—physical in the case of products or experimental in the case of services—provide the best evidence that a concept can be executed with the desired performance, including basic cost parameters.

---

[6] Testimony of U.S. Secretary of Defense, Donald H. Rumsfeld, before the Senate Armed Services Committee, Washington, D.C., May 16, 2002.

Marketplace acceptance. Unavoidably, some "unknowns" inevitably remain vague—most critically, the ultimate acceptance and success of the product or service in the marketplace. Although companies can usually predict costs fairly accurately, they can make only rough approximations of market acceptance until the product or service actually hits the market. Fred Smith, founder of Federal Express®, knew well the cost of his hub-and-spoke delivery service model, but he had only a gut feel for the likely response to this previously unavailable service concept.

At Disney's Imagineering, concept development tends to be the most critical stage requiring the greatest managerial judgment. Imagineers easily produce a host of ideas for new attractions in the ideation phase of Disney's process. Although, the creatives at Disney may not employ a harsh engineering term like "tollgate," the financial discipline effectively forces such decisions: substantial funding results only after Imagineering's customer, the park operator, agrees to fund further development of the concept. Then the Imagineers storyboard the concept in greater detail as a form of prototype while simultaneously developing a capital budget and revenue model that determines financial feasibility. At the completion of this phase, the Imagineers can predict with high accuracy the cost side of the financial equations. The revenue side, however, may remain largely unknowable until the attraction goes into actual service.

Brand and corporate image. Senior management also plays a critical role in safeguarding brand image during concept development. A seemingly viable project may simply become too much of a stretch for current consumers. Even worse, it could actually conflict with the desired brand image. Given Apple's strong industrial design skills—including an intuitive consumer interface and reputation for "cool"—the iPod, which stretched the company's computer brand into consumer electronics, proved readily feasible.

Volkswagen, however, apparently overstretched its brand considering the poor results from the 2003 introduction of the Toureg Sports Utility Vehicle, a high-end offering for VW. Jointly developed with Porsche®, which marketed its version as the Cayenne, the Toureg clearly had the performance to justify its premium pricing. But consumers could not extend the Volkswagen "people's car" brand to a $37,000 SUV. Porsche priced its version at a $5,000 premium to Volkswagen consistent with its higher brand image, but also faced some initial consumer resistance. For purists, a four door SUV hardly fit the Porsche sports car brand. The marketplace disagreed. Porsche sales jumped in the United States with the Cayenne selling more units than the 911, Boxster, and Carrera GT models combined in 2004. As these examples amply illustrate, senior management must guide project development teams to understand how far to stretch the brand.

Project metrics. Concept development should also result in a set of detailed project metrics for review by senior management at each successive tollgate. Some of the metrics obviously include investment budgets and timelines for project completion. Others might include market results such as sales volume, market share, margin goal realization, and product quality standards. Although teams may revise targets and metrics in later stages, they should always bear in mind that doing so may fundamentally alter the economic viability of the project. In addition, senior management should tie long-term metrics to long-term incentive bonuses of the team based on actual results achieved after product launch.

What new lessons did we learn during concept development?
What do we know about the market for this product? • *What did we learn from customer testing?* • *What customer insight is this product built upon?* • *Are customer needs and benefits clearly defined in terms of product specifications (function, performance, industrial design, etc)?* • *What is the potential market size that this product/service could serve?* • *What is the likely competitive reaction?* • *Does this concept fit with the channel?* • *Do we have channel buy-in?*
What are the implications for branding and corporate image? • *Does the product still support the brand/corporate image?* • *Is there an opportunity for sub-branding?*
What did we learn from the technical assessment? • *Are we technically capable of delivering the product?* • *Is the technical risk acceptable?* • *Have relevant trademark and patent searches been completed?* • *Has purchased/licensed product opportunity been considered?*
Does the preliminary business case support moving forward? • *How much value will this project likely create?* • *What are the risks associated with this project?* • *Do we have solid contingency plans?* • *What options exist if our primary path fails?* • *Is our advantage sustainable?*
What are the plans and resources required for design and build? • *What are the market, launch, and development plans and assumptions?* • *What portions of this project should be done internally versus externally?* • *Are all the appropriate functions involved and supporting the project as required?* • *Are small market trials recommended?* • *What resources are needed to get to the next tollgate (capital, expense, manpower from all required functions, special skill sets, etc.)?* • *Are the resource requirements in line for a project of this nature?*
On balance, should this project proceed to design and build?

Figure 4.6 Concept Selection Tollgate Questions

To manage the many risks inherent in concept development, senior management should challenge project teams as part of the concept selection tollgate with probing questions such as those presented in Figure 4.6.

Design and Build

* ✳ *

IDEATION	CONCEPT DEVELOPMENT	DESIGN & BUILD	LAUNCH

Because the design and build phase focuses on execution, it also demands a very different kind of executive and project leadership. The concept and market potential have been adequately validated. At this point, time, attention to detail, and quality become the defining factors for maximizing the profit potential of the product or service concept. As shown in Figure 4.7 in design and build, the team must complete the detailed design and testing of the product or service and prepare it for market launch, at the same time keeping on track and true to the project goals.

In this phase, the project team must turn all the hypothetical plans and assumptions into a real product or service. Theory steps back and reality rules. A robust plan leaves room for some unanticipated issues without fundamentally altering project economics. Nonetheless, project creation teams often face tradeoffs. For example, if some elements of the design exceed cost targets, should features be

Purpose of Design and Build Stage	Design and Build Stage Deliverables
Finish detailed production-ready design, test production process, prepare for market launch	• *Complete final business plan* • *Test product in all use environments* • *Qualify all production processes* • *Complete all market launch planning* • *Identify launch resources required*
Typical Launch Readiness Tollgate Issues	
• *Overall delays to design completion* • *Excessive cost overruns that threaten profitability* • *Unanticipated quality issues* • *Supplier failures* • *Operations/facility construction or installation delays*	

Figure 4.7 Design and Build Stage

dropped or should the cost target be revised? Project teams must make these difficult tradeoffs without compromising the project schedule and hopefully the potential return on investment. Senior management should provide clarity on the priorities at each tollgate.

In the mid 1990s Chrysler developed a new approach to target costing that greatly simplified the management of product creation project tradeoffs in procuring purchased parts from suppliers. Historically, company engineers would design the ideal part. The purchasing representative would then identify potential suppliers and negotiate the best price with one of them. To instill more financial discipline in the process, Chrysler took a new approach that firmly fixed the target cost for a part. The company then challenged suppliers to deliver as many of the features as possible for the given price. Some features were classified as "must haves," while others were "nice-to-have." Eventually, the supplier that could deliver the most "nice-to-haves" at the targeted cost won the work. The fixed target philosophy also applies well to other aspects of the product creation effort. Many projects have suffered from a "scope creep" that diminished the potential returns and, worse yet, failed to fully capture a profitable market opportunity because of missed deadlines.

What new lessons did we learn in the design & build phase?

Has the business plan changed based upon the detailed design?
- *Have any of the assumptions supporting the business plan (price, volume, cost, capital, timing, etc.) changed?*
- *How robust is the business plan and supporting assumptions?*
- *What are the remaining risks associated with this project?*
- *How has the current migration path changed from the original?*
- *Is the planned return worth the risks and investment compared to other viable options?*

Is the design finalized (product and production/service delivery process)?
- *Have "real" customers validated the design via actual usage?*
- *Are production systems and processes capable of delivering expected product quantity, quality and reliability?*
- *Are key suppliers' systems and processes capable of delivering expected quantity, quality and reliability?*
- *Has the product been tested in all environments where it will likely be used?*

Are the plans and resources required for launch ready?
- *Are supporting functional plans in place?*
- *How robust are the plans and assumptions?*
- *How much flexibility is available with these plans?*
- *Are contingency plans in place for areas where delays or problems could possibly occur?*
- *How will activities be monitored to ensure timelines are achieved?*
- *Is the total corporate structure in sync with the launch plans and execution activities?*
- *How does the launch of this project fit with other launches and the product portfolio strategy?*
- *What additional support will be required to execute these plans?*
- *Are the resources in line for a project of this nature?*
- *Are the resources budgeted in appropriate spending plans (capital, SG&A, etc.)*

On balance should this project proceed to launch?

Figure 4.8 Launch Readiness Tollgate Questions

The launch readiness tollgate at the end of the design and build stage confirms that the company can get the product out of the factory door or the service into the field; has primed the marketplace with advertising, merchandising, customer support, and service teams; and can deliver the product in sufficient mass to meet market demand. For example, at launch readiness for a software product at IBM, the team would have ensured not only extensive usability and reliability testing of the product, but also the company's ability and preparedness to sell, ship, and support the product in the field. A business-to-business firm like Otis elevators would have alerted architects, contractors, and construction companies of changes or new product features, trained the sales force, and ensured installation and support capabilities.

Figure 4.8 offers a set of representative questions that senior management could ask at the launch readiness tollgate.

Launch

* * *

IDEATION	CONCEPT DEVELOPMENT	DESIGN & BUILD	LAUNCH

The purpose of the launch phase is self evident. Figure 4.9 documents the purpose as well as this phase's deliverables and typical issues. One key to a successful launch is an appropriate ramp-up schedule. Slow ramp-ups lose valuable

PURPOSE OF LAUNCH STAGE	LAUNCH STAGE DELIVERABLES
Ramp up product or service delivery system to full-scale operation in a controlled manner	• *Execute low-rate production successfully* • *Perform full-rate production to plan* • *Meet or exceed product quality goals* • *Achieve customer satisfaction goals* • *Ensure product and customer support quality* • *Realize business plan objectives*
TYPICAL LESSONS LEARNED TOLLGATE ISSUES	
• *Failure to follow stage-gate process consistently* • *Missed milestones because of unexpected production or supplier issues* • *Sales miss projections due to faulty market or customer need assessment*	

Figure 4.9 Launch Stage

market opportunity, but ramping too fast can stress the operation and cause quality problems. Typical production ramp-ups last from three to six months, and even longer in a greenfield operation requiring significant shifts in staffing and training.

In addition to allowing employees to scale the learning curve, the ramp-up period allows time to fully verify quality and delivery performance across the supply chain. The ramp-up also provides time for training and implementing the required "backroom operations" like customer support and field service. Finally, the launch ramp-up ensures that the related market activities and challenges have also been addressed, such as point of purchase materials, advertising, and channel partner communication. Once the product or service is up to full capacity, the goal becomes the ongoing management of the product or service by adapting to input from the market and customers.

The final tollgate. Many stage-gate processes fail to provide for a tollgate at the end of the launch phase. Post-launch analysis, however, has shown that this final tollgate plays an absolutely critical role. The four-stage model presented here calls for a formal Lessons Learned tollgate that allows both senior management and team members the opportunity to review, assess, and reflect on what went well and what needs improvement. Leaders should capture and document these lessons, make them available to other product development teams and consider them for possible permanent change to the overall new product creation process.

Capturing and acting on lessons learned requires a discipline not always easy to achieve given the day-to-day pressures of completing the current project or the expediency of moving on to the next one. To mitigate these forces, a corporate product creation executive—the process owner—should take responsibility for capturing not only specific technical insight in design guidelines, but also for continuously improving the entire product creation process.

The process owner should also audit individual projects from time to time for process conformance. Lack of conformance tends to result from a few common root causes. First, the company culture may not have fully embraced the need for process discipline. If so, the process owner should bring the issue to senior management for reinforcement. More typically, however, a lack of conformance indicates insufficient training or a process that inadequately addresses the real life challenges inherent in the company's business model. The process owner should accept responsibility for addressing these root causes directly.

Continuous Improvement

* ✳ *

Companies should understand from the outset that even a well-designed process can, over time, become too rigid or simply out-of-step with current business realities. This does not necessarily mean that the process will change dramatically. Rather, it calls for improving the process continuously.

Applying the discipline of block upgrades, as discussed in detail in Chapter 5, the process owner should batch needed corrections and implement them all at one time every two or three years. This avoids the confusion that can result from continually introducing minor modifications. These block upgrades should have the formal approval of executive leadership and incorporate training to enable managers to support and act on them.

Improvement efforts should relate to more than process. They should also focus on areas such as training, staffing, and supplier support. Training current and new staff—both in-house and at key suppliers—in the best practices of new product creation can turn into a large undertaking for a company operating on a global scale. Such training usually entails detailed programs for full-time team members and condensed versions for part-time and supplier team management and staff. Training classes offer a great opportunity for senior management to appear before the participants, answer their questions, and show support for the corporate product creation process.

Although no two efforts will ever run exactly the same, the best companies consistently shorten the average length of their product creation projects while simultaneously improving financial and market performance. Often the impetus for improvement comes from the competition. In the 1990s the "Big Three" U.S. vehicle manufacturers and traditional European producers had to come to terms with a growing development gap compared to the Japanese. The U.S. and European companies took an average of five years to design a car while the Japanese accomplished the feat in four years. Over the ensuing decade, Western companies dramatically reduced development time to around two years. They still face a gap, albeit smaller, compared to the number one Japanese competitor, Toyota, which has improved its performance to about 12 months for a derivative design in Japan.[7] Obviously, no one can afford to stand still when shorter development cycles reduce costs and improve the odds of satisfying ever-evolving consumer tastes.

Engaging the Entire Organization

* * *

The preceding description makes clear that the product creation process requires strong support from the entire organization:

- Marketing helps translate the voice of the consumer.

- Engineering—or in the case of many service industries, the information technology department—drives the technical development.

---

[7] Ibid.

- Finance confirms and ultimately tracks the financials.

- Operations ensures the product or service can be delivered consistently in a "production" environment.

Depending on the company, different functions tend to take the lead in product creation projects. At aerospace companies like Pratt Whitney and Boeing, engineering drives the process. At consumer products companies like Bacardi of America and Mars, marketing tends to lead. Regardless of the leadership model, all functions provide critical support at particular junctures. And senior management must ensure they do just that.

The example for the hypothetical Omega Corporation depicted in Figure 4.10 shows the ebb and flow of a typical product creation project. Marketing and certain support functions play a relatively steady role throughout. Research engages early on but tapers at the end. The central ramp comes from the addition of engineers for the design and build phase. The operations function plays a role throughout but mostly towards the end. Although the functional and resource levels vary with every project, a continuously shifting pattern can be expected for every one.

Functional department support, however, goes beyond assigning the right people to product creation projects. Functional leaders must also ensure that those same people have a bright future once the project ends. Project performance critically depends on all corporate functions delivering support to the initiative. In many cases, project members reach back to their functional organization for additional support and special capabilities. The entire company needs to share in the goals, objectives, and excitement of the process of creating new products. This happens naturally when

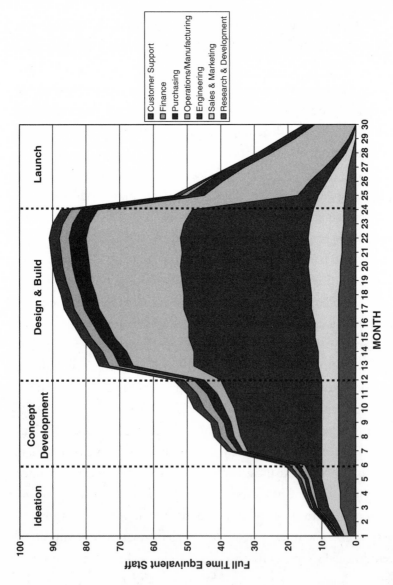

Figure 4.10 Omega Corporation Project Staffing Pattern

everyone at every level understands that they not only contribute to new product creation in some way, but that through the process the company works to ensure their future livelihood. By keeping everyone informed and involved in the new and exciting things on the horizon, executives energize their managers, who in turn, bring the spark to their own areas of responsibility. Quarterly reviews and "all hands" meetings offer an excellent forum for giving such visibility to project leaders and teams and for executives to demonstrate their commitment to innovation.

Engaging appropriately. Most executives focus their time proportionately to the level of staffing of the project. They spend little time during the front-end but get deeply involved as the launch date approaches. Unfortunately, though a natural reaction, such engagement typically proves disruptive rather than productive. In reality, senior executives should have much more influence on the project in the early stages, which require strategic insight and experience. At the later stages few decisions can be changed without incurring a huge cost and perhaps a marketplace penalty. Recognizing the needs of the stage-gate process, an ideal pattern of executive engagement should look more like Figure 4.11.

The figure shows a declining role for senior executives as the project evolves with major spikes around each tollgate and minor spikes of activity for monthly progress reviews. Senior management plays a critical role in making the "go/no go" decisions at the tollgates but should also be monitoring the monthly progress as well. Should a status report raise a red flag, the executive should reach for the phone or go visit the project leader to investigate in a way that leads the team to success without micromanaging. The figure also shows an important spike for the "lessons learned"

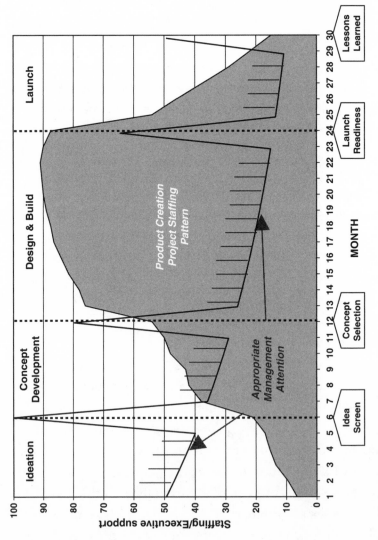

Figure 4.11 Senior Executive Involvement

tollgate. If senior executives don't personally demonstrate a commitment to learning from the past, the rest of the organization will certainly dismiss this important activity as well.

Senior executives should share responsibility for leading product creation projects. It should not simply be the responsibility of the functional lead of the development organization. For example, Whirlpool assigns one person from the executive committee, comprised of the top six to eight executives, to lend executive support to each major project. This executive keeps the committee informed between tollgates and milestones and acts on the committee's behalf to assist the project sponsor (typically a vice president but not a member of the executive committee) in removing any roadblocks that the team may encounter, such as financial or staffing constraints. In addition, the lead executive answers any questions that may require a strategic corporate perspective to prevent delays because of misunderstandings or lack of clarity.

The automotive industry has adopted a more formal model of shared responsibilities for functions and innovation. For example, Chrysler assigns product development responsibility around four product lines: Jeep®, minivan, small car, and large car. The executives running these four areas also have companywide responsibility for other functions including sales and marketing, procurement and supply, product strategy, and external and regulatory affairs.

Although necessary, a well-designed stage-gate process with appropriate senior executive involvement is not sufficient. The people assigned to product creation remain the most important factor. Do they have the right incentives? Is there an appropriate mix of creativity and pragmatism? Does the culture and environment foster innovation?

The successful execution of the product creation process forges the key links among understanding the voice of the customer, satisfying customer needs, and delivering the product portfolio. Chapter 5 explores how this all ties together to support building and maintaining a competitive product portfolio. Later, Chapter 6 will focus on key elements of people and project management needed to inspire and lead the workforce.

THE COMPETITIVE PRODUCT PORTFOLIO

* * *

Integrating Process and Product Architecture

Viewed as the father of the modern corporation, Alfred P. Sloan led General Motors Corporation from 1923 to 1946. He also played a seminal role in building the foundation for modern product management. In the company's 1924 annual report, Sloan articulated his vision of GM providing "a car for every purse and purpose." His decision to pit his full and varied product line against the then dominant but narrowly focused Ford Motor Company represented more than a marketing strategy.

His vision came from what Sloan described as the "irrationality" of the General Motors line-up in 1921 and his decision to develop a corporate product policy. At that time, the company offered ten different cars under seven different nameplates: Chevrolet, Oakland, Olds, Scripps-Booth, Sheridan, Buick, and Cadillac. Without clear market positions, each car and brand competed with the others. Except for Cadillac and Buick, all of the brands were losing money.

The original product policy, the development of which Sloan considered to be "one of the most significant in the evolution of the company,"[1] proposed a strict product line structure. Over time the company aligned its brands against explicit price ranges extending from the low-end competing with Ford to the high-end, which Cadillac came to dominate. The clear policy allowed the company's divisions to show the entrepreneurial spirit necessary in a dynamic market while ensuring a common direction overall. Although Ford had a commanding 60 percent share of the total unit sales of cars and trucks in 1921, General Motors by the beginning of the next decade gained market leadership, which they maintained for more than 70 years.

Sloan also introduced a centralized purchasing committee to capture synergies across the decentralized divisions. Although he admitted in his autobiography, *My Years with General Motors*, that the effort could not "be cited as an unqualified success," it did drive the company to "standardize articles where possible." Sloan also created a central engineering staff to focus on problems common to all of the GM's divisions. This push for centralization and standardization ultimately led to the concept of a platform architecture adopted throughout the automotive industry in the 1970s and 1980s.

Maintaining a competitive product portfolio—as Alfred P. Sloan envisioned—requires integrating processes and product architecture. Executives must ensure that product management teams, product creation teams, and researchers systematically integrate their efforts to ensure that market-back needs align with

---

[1] Sloan, Alfred P., Jr, *My Years with General Motors*, Doubleday, 1963.

technology-forward possibilities. Equally important, the product architecture must efficiently support the ongoing product changes that keep the product fresh and vibrant.

Integrating Innovation Portfolios

✳ ✳ ✳

Executive management plays a critical role in ensuring that the company's portfolio of products or services remains responsive, relevant, and competitive in the marketplace. Success in doing so requires integrating all three of the company's product or service innovation portfolios:

1. The current portfolio of products or services (sometimes referred to here more simply as the product portfolio)

2. The portfolio of advanced technology capabilities

3. The portfolio of product creation projects

This is no easy task. Each of these portfolios—product, advanced technology, and product creation—has its own set of stimuli and timelines:

- The product portfolio must remain responsive to customer feedback and the competitive environment on an ongoing basis. As critical as this market-back approach may be, it can also place excess emphasis on marginal enhancements and "me too" products.

- The advanced technology capabilities portfolio can focus on significant new capabilities, leading-edge innovation, or adapting an existing technology to a new purpose. Its technology-forward approach, by definition, puts it on an unpredictable timeline.

- The product creation projects portfolio, while adhering to the discipline of the stage-gate process, must operate on a relatively tight timeline to assure that new products arrive in the marketplace at the right time.

Effective execution of the company's product strategy requires integrating these three portfolios. Such integration also requires a clear product or service architecture and a systematic management process. Because of the inherent differences among the three portfolios, however, disconnects in focus, timing, and priorities can occur. To achieve optimal business results and to remain competitive, senior management must minimize such disconnects. This means integrating the technology-forward approach characteristic of research and development with the market-back approach of sales and marketing.

Creating an advanced technology capability portfolio entails the most uncertainty. A hard link exists between the product creation portfolio and the product portfolio. A softer link exists between the technology capability portfolio and the product creation portfolio. This link dictates the need for contingency planning in product creation projects because a technology may not always be verified in time to be included in a particular product creation project. Mars finds this particularly important in linking advanced process technology development to product creation. The head of innovation engineering has responsibility for deciding when to plug new process technology into the innovation process and develops contingency plans, like outsourcing, to cover the risks. Systematic nurturing—and pruning—of each of the three portfolios is also necessary to assure that they all maintain the right focus and mix for the market.

Leading companies accomplish this by using cross-functional product management teams (PMTs) to identify the issues and to facilitate portfolio linkages and sound decision-making. When done well, all involved will have a clear understanding of the integration of advanced technology capabilities, product creation projects, and the product/service portfolio. Figure 5.1 offers a highly simplified illustration of how advanced technology capabilities feed into product creation projects, which then link to the product portfolio. Maximizing the breadth of a competitive product portfolio at a reasonable cost, however, demands a clear product/service architecture as well as a systematic management process.

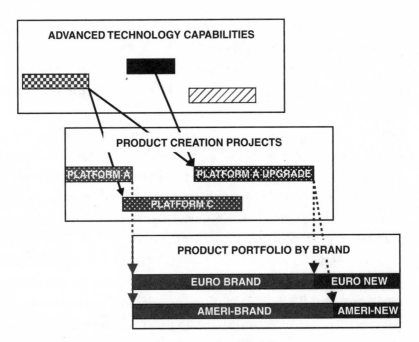

Figure 5.1 Integrating Product Innovation Portfolios

Product Architecture

* ** *

A company's product or service architecture comprises the "building blocks" for managing the product portfolio. By leveraging common building blocks and systematically mixing, matching, and upgrading others, a company can offer a wide array of competitive products or services without designing each and every one from scratch. The primary building blocks include platforms, modules, and differentiators.

Platforms provide the basis for a number of models that serve different market segments but share some underlying common features. Modules help reduce part counts and manage complexity. Differentiators result from a clear understanding of what matters to the customer and what can or should remain invisible. Thoughtful management of product architecture using platforms, modules, and differentiators can deliver efficiencies in product creation, purchasing, manufacturing, and service while offering superior customer benefits.

Platforms. Returning to the example of General Motors, the company's worldwide model lineup includes about a dozen platform variants. For example, the Chevrolet Malibu, the Pontiac G6, the Saab 9-3 and the Opel Vectra all share the same midsize, front-wheel-drive platform. Similarly a single rear-wheel drive platform supports the Buick Rainer, Chevrolet Trailblazer, the GMC Envoy, and the Saab 9-7X. Although these models build on a common chassis and suspension design, each offers a feature bundle and price point consistent with the supporting nameplates— Buick, Chevrolet, GMC, or Saab. The different combinations of platforms among the variety of GM nameplates provide a unique lineup for each brand despite a high degree of commonality across all product lines.

Farm equipment manufacturers tend to define a common industry architecture for tractors based on engine power: compact tractors (20-40 horsepower), utility tractors (40-80 horsepower), row crop tractors (80-130 horsepower), and large tractors (130-250 horsepower). Manufacturers combine these platforms with different attachments such as planters, mowers, front-end loaders, cultivation equipment, and backhoes to deliver a wide range of functionality. For larger, specialized equipment, the manufacturers tend to create functional platforms such as forage harvesters, combines, and cotton pickers. Over time, companies retire certain platforms and replace them with a new structural design that serves as the basis for future product design families.

Platform architecture allows the product manager to position multiple brands and feature bundles in different distribution channels or markets with minimal incremental cost. Product developers often forget the variety of derivative product iterations and focus instead on the platform only. To avoid this myopia, they should view the product from the customer's perspective. Such "fresh eyes" can often reveal opportunities for competitive differentiation.

Although widely practiced and generally visible in industries such as automotive, commercial aircraft, and other large physical products, the platform concept is less obvious in industries like food, software, and financial services. Yet the benefits of defining platforms and strategies in these softer product industries can prove just as valuable and lead to product extensions and efficiencies as well.

For example, Mars applies platform logic to its products with a linkage to its manufacturing technologies. The company uses its proprietary process technology for hard shell coating across a variety of softer inner substances beyond the chocolate used in traditional M&M'S. The technology has been applied across a variety of chocolate-coated products including crisped rice and peanut butter for M&M'S derivatives, and the chewy center of Skittles®. In general, product managers find that the platform is the most powerful organizational tool for ensuring order and efficiency in the product lines. All companies should think creatively about their definitions of product platforms and consider both physical and virtual platform models for product architecture.

Modules. A modular design philosophy complements the platform architecture by simplifying the process of creating variations of the common platform. In its extreme form, as in the personal computer industry, individual consumers can customize the product to their own specifications. For example, thanks to a modular architecture, Dell can offer a wide range of configurations for its Inspiron 6000™ laptop computer. Standard configurations in 2005 included the base system for $999, a music and photos arrangement for $1560, the ultimate productivity set-up for $1885 and the ultimate entertainment configuration for $2217. These variations derive from modular combinations of different Pentium® processors, varying amounts of memory, and various size hard drives. Further customization of the hardware can increase the price another $919 by enhancing the processor speed, upgrading the monitor, adding a sound card, and boosting the battery life.[2]

Modular design may also simplify the block upgrade process. A block upgrade involves aggregating several new features to keep the product fresh and interesting while avoiding the cost and effort of a major redesign. Through periodic and systematic block upgrades, companies can present an orderly picture to the market through advertising and point of purchase materials. Without a modular architecture and block upgrades, development teams

---

[2] Based on Dell.com consumer website offering, November 2005.

sometimes delay product creation projects in an attempt to get every last feature possible into the current version under development. But with modular designs and block upgrades, ideas that surface well into the current development cycle can be slated for inclusion in a module in the first block upgrade after product launch.

Modular designs can also reduce the number of parts required and can simplify outsourcing of major subsystems. The automotive industry has greatly increased its use of modular design since the early 1990s. Seats now come in modular designs including integrated seat belts and other safety and convenience features. The Mercedes-Benz designed its M-class, built in Tuscaloosa, Alabama, with a cockpit module that includes a dashboard, instrument cluster, steering column, pedal assembly, airbags, comfort control equipment, and venting. Outsourced to Delphi-Packard, the module—which consists of 150 individual parts from 35 different suppliers—comes directly to the assembly line in sequence order to match the color and options package of the specific vehicle.[3]

Modular designs, appealing as they may be, also have a downside. They can force tradeoffs and potentially a suboptimal design with the need to create and maintain clear interfaces with existing modules. For example, today most office copiers reflect modular designs with different sorter and paper tray options depending on the customer's needs. A more integral design would lower the cost of the unit in isolation, but raise the cost across the full product line in aggregate. In fact, the low-end versions of

---

[3] Hoffman, Kurt C., "Mercedes' Finely-Tuned Supply Chain," *Global Logistics & Supply Chain Strategies*, May 1997.

most products do not employ modular designs as a way to avoid any unnecessary interface costs.

Although few actively manage these issues well, companies can gain significant savings from explicitly balancing the tension between customization and standardization. For example, a car manufacturer could design a single jack that would operate across a range of models. If they designed it to lift a two-ton sports utility vehicle, however, it would be significantly over-engineered for a small car. In the mid-1990s, Ford launched a major project to reduce complexity after discovering that it had sixteen different types of grey trunk carpeting across its model lineup. Ironically, the moderately priced Ford Escort had a more expensive grade of trunk carpeting than the luxury-priced Lincoln Towncar.

Companies can also manage design tradeoffs by using virtual rather than physical module definitions. The current automotive industry approach to the design of brake systems offers a good example. Rather than designing the calipers, the master cylinders, and control systems independently, most vehicle manufacturers assign full brake system responsibility to a single supplier. Even though the products don't arrive at the plant as a single module, the design leverages opportunities for improved performance at the systems level.

Despite the advantages gained from reduced complexity, excessively modular product architectures can lead to turning the end product into a commodity. When modules become completely interchangeable within an industry, the end-product assembler loses its market power and the suppliers of competitively advantaged components begin to accrue the majority of industry profits. Charles Fine of the Sloan School of Business at the Massachusetts Institute of Technology documented this phenomenon in a variety of industries including computer hardware, media and entertainment, and bicycle manufacturing.[4]

This phenomenon can be seen quite clearly in the personal computer industry where none of the major manufacturers except Dell makes any significant profits, while the "component" suppliers like Microsoft and Intel reap large returns. Dell's profitability does not serve as a counter-example of the trend, but instead highlights that even in a thin margin industry a competitor with a uniquely efficient operating model can succeed. Fine also argues that industries with excessively modular designs will ultimately revert to an integral design. Modular designs force tradeoffs to keep the interfaces between modules simple, but at some point, an integral design will offer a substantially superior tradeoff and obsolete the modular advantages.

Differentiators. In most industries, companies retain the ability to further leverage the benefits of modular design. Successful adopters of modular design identify and continually innovate a set of differentiators while standardizing modules and employing platform architecture to reduce cost in areas of little importance to the consumer. Trunk carpeting would not be a substantive differentiator in the purchase decision of many consumers. Ford, therefore, could choose to standardize its options to one or two, achieve lower cost overall, and not have an appreciable effect on consumer choice.

Whirlpool uses the colorful term, the "green line," to highlight the distinction between differentiators and less critical product attributes. Features that consumers perceive and value reside

---

[4] Fine, Charles H, *Clockspeed: Winning Industry Control in the Age of Temporal Advantage*, Perseus Books, Reading Massachusetts, 1998.

in front of this imaginary green line. Those that customers do not perceive or value lie behind. In a refrigerator for example, the cooling system, insulation, and control system lie behind the green line. The aesthetic design, including interior shelves, drawers, ice maker, and control features remain in front of it. Whirlpool encourages product teams to leverage existing platforms and modules to control costs behind the green line and to focus creative efforts on new features in front of the line that create competitive differentiation perceived by the consumer.

Although implied by the Whirlpool example, differentiators need not be physically visible, or even consistent among competitors and market segments. The suspension and power train performance serve as differentiators in high-end sports cars like a BMW or Porsche. Conversely, they clearly fall behind the green line in most mid-sized family sedans. Family car buyers put more emphasis on practical functional features like interior seating design, fuel efficiency, and safety systems. A sports car enthusiast may be quite satisfied with a minimalist interior as long as the vehicle accelerates quickly and handles well at high speeds.

The green line concept applies equally well in a "service" company. Differentiated reservation services provided to frequent flyers by major airlines offer the most valuable travelers a special toll-free reservation number. Call center staff greet the preferred customers by name and route them to the front of the queue. Unknown to most customers, a single call center—well situated behind the green line—provides service to all callers creating staffing efficiencies without forcing the most important customers to wait longer than necessary for a higher level of service.

Features can also lose their differentiating capacity over time. Disc brakes, power windows, remote entry locks, and cruise control, all significant differentiators when first introduced have over time migrated behind the line as standard options on a wide range of vehicles. The Intel Pentium processor offers a case of reverse green line migration. Intel moved its microprocessor into the consumer's field of view by offering a price discount to computer manufacturers willing to add the "Intel Inside" sticker on the outside of the end product. With their razor-thin margins, manufacturers grabbed the short-term gain—but ultimately became further locked into Intel. Although other microprocessor designs emerged from smaller competitors, Intel's direct consumer advertising created the perception among consumers that computers with anything other than an Intel microprocessor were inferior. Whirlpool, which manufacturers most of the Kenmore® brand laundry product line for Sears, uses a similar component branding strategy with its Catalyst® wash system.

Product and Service Portfolio Management

✳ ✳ ✳

Although the development of new technologies and products seems to take center stage here, in reality, effective product creation often involves making ongoing minor adjustments to the existing product or service portfolio. These adjustments can be as varied as special promotional packaging for consumer goods to introducing new color and trim options for an automobile. Mars, for example, rolls out special colors and packaging for its candies for holidays ranging from St. Valentine's Day to Independence Day to Halloween to Christmas and also special packaging for certain retailers like Wal-Mart. Such changes keep the product fresh in consumers' eyes and help keep key competitors at bay.

BMW initially introduced its Z4 Roadster in 2003 with the choice of only three exterior colors—silver, gray, and black—with a black convertible top and black leather interior. Two years later the offering had expanded to eight exterior colors, three different convertible tops, and three colors of leather seats. Such changes help maintain consumer interest beyond the initial introduction, critical for the Z4 to sustain a seven-year model life like the Z3 Roadster, which preceded it.

Product management teams (PMTs) use a variety of short-term and tactical activities to maintain the health of their product or service portfolios in the marketplace. PMTs must address market reaction to their product, defend the current competitive position, and respond to changes in customer buying patterns. They often do so by delivering new product versions with minor variations in functions, benefits, and pricing. PMTs also weed out low-volume and low-margin items and work on cost reduction and quality improvement programs. Not surprising, PMTs also serve as a breeding ground for new product ideas and product improvements. However, their front-line input often tends to be tactical in nature and to point towards evolutionary product improvements rather than radical enhancements and true innovation.

As shown in the simplified schematic in Figure 5.2, a company's product or service portfolio evolves over time through the introduction of new platforms, major updates of new platforms, and simple block upgrades to existing platforms. In our simplified example, the hypothetical Omega Corporation competes in two

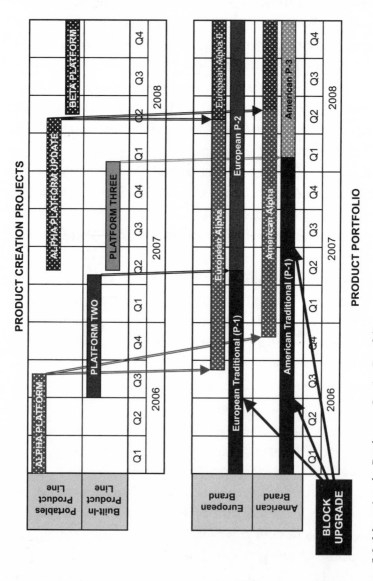

Figure 5.2 Managing the Product or Service Portfolio

market segments: portables and built-ins, and in two geographical regions, Europe and North America. Currently Omega offers a traditional product for the built-in segment under a different brand name in the two regions. For simplicity, the figure refers to the current offerings as European Traditional and American Traditional.

As shown in the lower left-hand corner of the chart, the company plans a "block upgrade" at the beginning of the third quarter of 2006 in both its European and American regions. Recall that a block upgrade involves aggregating several new features to keep the product fresh and interesting while avoiding the cost and effort of a major redesign. A block upgrade can encompass substantial changes to the functions and features of the original platform, which can apply to any or all products based on the platform.

Lacking a block upgrade strategy, companies tend to take one of two non-productive routes. Some companies simply release a continual stream of new product features, which adds an element of confusion for customers not to mention everyone else involved in production and product support.

Other companies delay new product introductions in an attempt to incorporate new ideas, (often suggested late in the process by the sales and marketing department) under the assumption that they have only one chance to incorporate the new idea. With an organized block upgrade strategy, however, the product team can introduce new ideas with a coherent marketing message in an orderly way without impeding existing product creation projects. Block upgrades also minimize chaos and confusion for areas such as production, the supply base, customer support, and field service.

Returning to the hypothetical Omega Corporation, although it currently only offers a built-in product line, it has a product creation project underway to introduce a completely new platform in the portables market, the "Alpha Platform." With this product creation project scheduled for completion in the third quarter of 2006, the product portfolio plan calls for an initial introduction of the European brand followed one quarter later by an American branded version.

Although the Alpha platform is completely new, the company has already scheduled the launch of an upgrade project for it less than a year after the initial launch. The company pursues such a plan for several reasons. A pre-planned major upgrade deflects the natural inclination toward "scope creep" in the initial product creation project. It allows the initial project team to remain committed to the original introduction date. It also reflects the inherent need to continuously renew any product offering: successful companies view product creation as an ongoing process rather than a "one-off" event of invention. More likely than not, this proactive approach will position the company in an offensive rather than defensive stance in the marketplace.

Omega Corporation also has future plans for its built-in product line. It has slated two new platforms for introduction in 2007 and 2008. These will replace the original platform used for both the European and American versions of the traditional built-in product line. The product management team has determined that the European and American markets have begun to deviate from one another. The team also believes that two distinct products built on different platforms offer the best option for meeting the divergent customer needs.

As reflected in the figure, the Platform Two product creation project will develop a replacement for the traditional built-in product in Europe, but the American built-in will be replaced by the Platform Three project in 2008. Also note that given the later introduction of the new platform in America, the company will implement another block upgrade to the American built-in in the third quarter of 2007. Such an approach spreads the investment cost of the new platforms over time by extending the life of the current product in at least one market.

Even this simplified, hypothetical example—two product categories and two brands and two regions—presents significant complexity. Imagine the intricacy of a real company such as Whirlpool or General Motors. Whirlpool sells ten major categories of appliances: washers, dryers, refrigerators, dishwashers, stovetops, ovens, freezers, microwaves, trash compactors, and room air conditioners. It also blurs traditional category boundaries—as in the Polara® refrigerator/oven—and even introduces entirely new categories of major appliances like the Portable Valet® clothes refresher system. The company also sells its appliances throughout the world under a variety of global and regional brand names including Whirlpool, KitchenAid®, Brastemp®, Bauknecht®, Consul®, Maytag® Jenn-Air®.

Alfred Sloan's General Motors continues to deal with the challenges of complexity. Today the company offers more than fifty different models of trucks, SUVs, vans, sedans, and sports cars under eight brands in the United States alone—Buick, Cadillac, Chevrolet, GMC®, Hummer®, Pontiac, Saab, and Saturn. Outside of the United States, General Motors also has its Holden, Opel, and Vauxhall operating units. Adding even more complexity, GM also coordinates product offerings with investment partners Isuzu, Suzuki®, and Daewoo.

Advanced Technology Capability Projects

* * *

While the product management teams work "market-back" to manage block upgrades and to time the introduction of new products, research and development programs focus on a "technology-forward" approach as discussed in Chapter 3. Core technologies are created and developed in programs addressing strategic research priorities in a particular technology domain. These programs validate concepts in real or simulated environments on advanced technology demonstrators (ATDs) with functioning prototypes. Once validated, the technology or innovation may be considered for a product creation application or for direct introduction in a small block upgrade. Teams may also use ATDs to assess new developments brought forward by suppliers or other outside sources. Some new ideas require very little technical development and can quickly move from concept to market introduction. A substantially new technology, however, should always be validated on an ATD before widespread market release.

Just as product creation projects must be mapped to the product/service portfolio, so too must advanced technology capability projects map to product creation projects. As shown in Figure 5.3, a continuation of the hypothetical example created

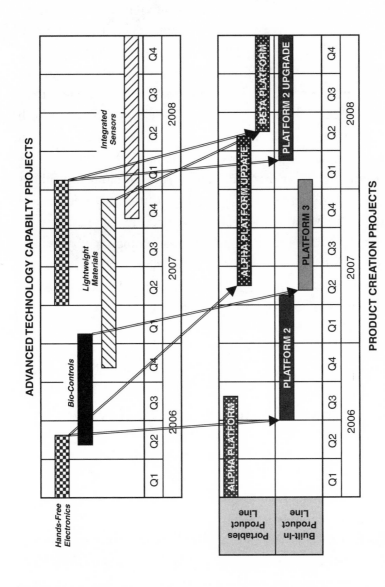

Figure 5.3 Integrating Advanced Technology Capabilities

previously, advanced technology capability projects can take many forms. The hypothetical Omega Corporation plans to leverage its research into hands-free electronics in both its built-in and portables product lines. The European platform will employ hands-free electronics, while the American platform will leverage a different technology, bio-controls. Although Omega will leverage certain technology capabilities, such as hands-free electronics, across both product families, it will focus its advanced technology efforts in lightweight materials exclusively on the portables segment. The figure also highlights that some advanced technology capabilities represent ongoing strategic research priorities, with continued upgrades, for example the repeated projects for hands-free electronics, each intended to push the technology a bit further. Finally, if the advanced technology capabilities shown in the figure are not validated in a timely manner Omega has a contingency plan to assure the timely introduction of the platform.

Annual Product Strategy Review

* ⁂ *

Senior management can ensure the integration of the various advanced technology projects, product creation projects, and product portfolios through annual product strategy reviews. Although rare even among leading companies, the annual product strategy review ranks equal in importance to the more common processes of strategic planning and annual budgeting. Many companies conduct a multi-year strategic review at the beginning of the fiscal year and then finalize the next year's operating budgets near the end of the fiscal year. An annual product review held between these two planning events can focus on the real lifeblood of any growing business: the ongoing nurturing of the product or service portfolio.

Building on the company's understanding of customer trends and the anticipated strategic moves of major competitors, the annual product review allows management to set priorities and to integrate development efforts across the full spectrum of project and product portfolios as shown in Figure 5.1. Once set, these priorities inform the specific annual operating plan, which allocates expense and capital budgets based on product creation commitments.

Product strategy reviews differ significantly from corporate strategy reviews. The latter tend to focus on strategic direction, new lines of business, revenue and profit goals, and other opportunities such as mergers and acquisitions. A product strategy review, on the other hand, forces senior management and the product management teams to assess the total product offering both as it exists currently and as it will evolve in the years ahead.

Preparing for the product strategy review serves as a good forcing function for all product areas requiring them to communicate their plans and rationale in a comprehensive way without the distraction of other business issues. Product strategy reviews typically bring to the surface "holes" and opportunities in the product portfolio. They also serve as a forum for ensuring a common understanding of major customer trends across the product portfolio and developing answers to key questions that cross product creation boundaries. Should one product type be expanded more rapidly than another? Are the technology concepts being prioritized to the best consumer applications? How are competitive product portfolios changing?

Senior management can use the product strategy review to tighten the linkages among the advanced technology groups in research and development and the various product management teams. The process forces all to look outside to calibrate the typically internal focus on the next "new, new thing" or the most recent minor product upgrade. The review starts with an outside-in marketplace perspective on customers and competitors and then works "market-back," through Figure 5.1. Starting with that same outside-in perspective, the review assesses the entire portfolio "technology-forward," as well. This systematic analysis typically identifies areas where senior management will clarify and often alter their product portfolio objectives.

Ultimately, the collective group of senior management, product management teams, and technology managers hammer out the priorities to ensure that the various portfolios of advanced technology and product creation projects will generate the most competitive product or service portfolio possible and as fast as possible. The size of the investment will dictate the pace of change, which suggests a natural transition to the annual budgeting process. The portfolio makes sense when market-back and technology-forward reviews align at an investment level the company can afford. The product strategy review then feeds the budgeting process in an informed manner so that the new annual budgets become objective-driven rather than merely an extrapolation of past revenues, margins, and cost.

Keep the Big Picture in View

* * *

The integration of a systematic process of product strategy review and a clearly articulated product or service architecture allows a company to drive continuous innovation while leveraging scale economies. Senior management plays a crucial role in setting the context for achieving the right balance. Individual product mangers, designers, engineers, and researchers generally prefer to "design from scratch" with a primary focus on their own piece of the big picture. Senior management should maintain the process discipline necessary to ensure the linkages and the strategic vision required to guide the architecture investments. Chapter 6 offers further guidance on the necessary soft skills for managing people involved in product creation as well as the harder skills of project management.

Chapter 6

PEOPLE and PROJECT MANAGEMENT

————— ✳ ✳ ✳ —————

Inspiring and Leading the Workforce

Walt Disney not only mastered his own creative craft, he also inspired extraordinary and wide-ranging creativity throughout his far-flung enterprise. In 1952, when Disney envisioned a family theme park based on his forte, storytelling and animation, he began by recruiting a hand-picked team of some of his studio's best writers, directors, and artists—the creative force that had helped make the Disney name a household word. With that small core group under his inspiring leadership, Walt Disney designed, engineered, built, and marketed Disneyland. His dream of a magical kingdom came true.

Today, that original team has evolved into the Disney Company's creative think-tank and development subsidiary, Walt Disney Imagineering. Appropriately, the company's name combines the two elements—imagination and engineering—that Walt Disney knew he needed to realize his product creation vision. Today, Walt Disney Imagineering employs more than 1000 "cast members" from more than 140 disciplines including model makers, software developers, artists, sculptors, writers, engineers, architects, music experts, special effects designers, project managers, film makers, scientists, animators, and landscapers to name only a few.

Unlike Walt Disney, few executives today can claim to have mastered the creative disciplines they must manage. Rather than leading purely by personal example, most executives must focus on building and nurturing the appropriate product creation culture. Don Goodman, current president of Disney Imagineering, for example is a former certified public accountant—a profession that traditionally has not always viewed creativity as a virtue.

If Walt Disney could be characterized as the conductor of a great symphonic orchestra, Don Goodman would liken his own current role to that of the orchestra's business manager. The business manager must recognize that the performers simply want to make great music. Even though they care less about business details than their art, the musicians also understand that the symphony must attract an audience and make money to survive. The business manager therefore must instill the proper discipline within the group to ensure the right end result and at the same time nurture the artistic spirit that makes the product great.

Create the Right Culture

* * *

Leading the people involved in corporate product creation can seem extraordinarily difficult. It can also prove extremely rewarding. The people who gravitate toward innovation activities tend to be very smart, highly creative, and more motivated by recognition and the freedom to create than by the money they make. To lead and inspire such a workforce, senior management must find a delicate balance between seemingly contradictory forces: top-down direction versus individual empowerment; experienced judgment versus creative license; pressure to perform to expectation versus willingness to challenge convention; and by-the-book execution versus pragmatic adaptation. Achieving the right balance among such forces can give a company a true competitive advantage.

Attract the Best

✳ ✳ ✳

The intangible value of a strong product creation capability comes not from the manuals or processes, but from the embedded knowledge and experience of those responsible for making it happen. The most brilliant scientists, engineers, and researchers often prefer to work alone, doggedly pursuing a personal passion. Ironically, most research projects, and certainly all product creation projects, require teams of individuals with diverse skills working together toward a common goal.

Disney Imagineering, for example, employs story tellers, artists, engineers, set designers, and robotics experts to make its "magic." Whirlpool integrates marketing, engineering, procurement, manufacturing, finance, customer service, and software experts to maintain its global leadership position. Mars engages chefs, marketers, packaging experts, ethnographic researchers, process engineers, and financial specialists to sustain consumer-driven innovation.

The interaction of individuals with diverse backgrounds expands the perspective of the organization and therefore its output. Senior management needs to find and recruit intelligent, creative people—experts in their specialty who also have the temperament to work collaboratively and cooperatively. In addition to their demonstrated intelligence and creativity, these individuals must also be open to ideas that are not their own. They also should not be constrained by paradigms, but willing to challenge conventional wisdom. Finally, they should be self-motivated and comfortable with ambiguity. Brilliant individuals who think they have all the answers all the time ultimately stifle the organization, even if they are "right" all the time.

Andrew Hargadon from University of California Davis asserts that an individual's "social capital" could prove as valuable as their "intellectual capital." His research shows the fallacy of the "lone genius" inventor. Hargadon notes that even the prolific Thomas Edison had a key technical partner in Charles Batchelor, an Englishman "whose training as both a mechanic and a draftsman complemented (and grounded) Edison's more flighty visions."[1] Edison's greatest contribution may well have been the vast array of contacts he had in the nascent telegraph, electricity, and railroad industries.

---

[1] Hargadon, Andrew, *How Breakthroughs Happen: The Surprising Truth About How Companies Innovate,* Harvard Business School Press, 2003.

Allow Freedom in Context

* ✳ *

Intelligent, self-motivated people generally desire—and deserve—considerable freedom to deliver their best work. This proves particularly true for the creative individuals drawn to research. 3M pioneered an explicit policy that allows its staff members to devote up to 15 percent of their time to discretionary projects of their own choosing. The policy encourages freedom, but puts it in context: 85 percent of a person's time will be explicitly aligned with corporate objectives, underscoring the company's understanding of the need to find the appropriate balance between creative freedom and project discipline.

Senior management should make sure that researchers and product creation staff participate in activities that connect them to the real world to help balance their instinctual aspiration for complete creative freedom and to ensure that their creativity is channeled into productive efforts. To stay grounded, researchers should regularly spend time on the more pragmatic, application activities of a product creation team. Engineers, normally deployed to product creation teams, should also spend time in the manufacturing plant or on service calls observing the customer first-hand. The need for rotating individuals beyond product creation roles applies to the business functionaries as well. Procurement people supporting product creation should not focus exclusively on advanced technology efforts: they should experience the pressures of long-term contracting and annual price negotiations as well as the excitement of innovation. Financial specialists should develop an appreciation of field operations in addition to their focus on the staff activities at corporate.

Foster Openness

* ✳ *

Obfuscation has dealt the death blow to many product creation projects. Individuals and project teams may hide or downplay problems, not out of dishonesty, but because they sincerely believe they can find solutions without management intervention.

Despite a team's best efforts, problems may surface late in the product creation process that could—and should—have been solved sooner and more effectively by making senior management aware of them earlier. A team can also waste precious resources and time trying to fix a problem that senior management might have solved by simply changing some of the project objectives. Senior management must ensure that teams understand that bringing problems to the surface is expected. Identifying problems earlier is best, but even late in the process is better than enduring a debilitating crisis after product launch or service rollout.

Senior management must instill an environment that encourages people to bring critical issues or problems to the forefront as soon as they spot them. Management can also look for the subtle clues of potential problems—body language or an off-hand comment, for example. Another potential pitfall is for team members to "cover" themselves by documenting problems in overly technical language, footnotes, or addenda, rather than highlighting them.

Part of creating a culture of openness comes from the ability to spot such tell-tale signs. As soon as they suspect a problem, managers should drill down to the truth. Just as important, active engagement with the teams and a reputation for not "shooting the messenger" give team members the confidence to highlight problems. One important caveat: leadership means maintaining a delicate balance. Senior management must avoid becoming the solution to every problem. When teams feel empowered and accountable for solving problems without elevating everything to senior management, they are more likely to understand when and when not to raise the red flag.

Lead, Don't Micromanage

* ✳ *

The most effective executives *lead* the product creation teams rather than *micromanage* them. They challenge and provoke their teams to consider and adopt new perspectives, not simply because a senior manager offered the perspective, but because of a compelling reason behind the executive's point of view. Some executives, like Lee Iacocca or Walt Disney, have an instinctual knowledge of the market and the creative process. Successful executives have a depth and breadth of experience far beyond that of the average team member. They should share their experience—in a constructive, non-threatening way—to generate the best results for their organization.

Mandating from on high can produce malevolent conformance: teams resolve to follow an idea to its extreme just to "prove the executive suite wrong." Rarely, if ever, should a team feel obligated to move in direct opposition to its shared perspective. Issues of the appropriate project direction should be resolved by respectful debate, or if absolutely necessary, by restructuring the team. Whether consciously or subconsciously, only rarely will a team do a good job at implementing a concept that is at odds with the teams' own firmly held beliefs.

Learn from Failure

* ✳ *

In addition to effective, ongoing interaction with product creation staff, senior management can foster a culture of disciplined creativity by managing failure effectively. Legitimate reasons for "failure" abound. Market conditions can change; ideas that offer a great consumer proposition become financially unattractive; great concepts prove technologically infeasible; or an idea is simply ahead of its time.

The project teams—as well as staff throughout the company—should understand and accept that terminating a project is a *desired* outcome in some cases. If all ideas progress through every review process and formal tollgate without a hitch, the process is not working properly. A lack of failures may indicate that unacceptable projects have advanced through the product creation process because of momentum. It could also mean that the company has not considered enough creative, potentially high-payoff projects at the front end of the pipeline.

Although difficult for individual team members to accept, a pattern of no failures suggests very little true innovation for the company as a whole. Too great a fear of failure can also stifle the type of open communication needed for effective new product and service creation. Senior management must signal the acceptability of a certain level of failure by ensuring that members of terminated projects move on to other promising positions within the company. This should be made clear upfront always and delivered upon consistently.

If management has truly staffed the organization with the "best and brightest," finding desirable positions for product creation team members—whether after a failure or a success—should not be difficult. If troubles arise in placing them, the root problem could be poor support from the functional organizations. Functional managers should enthusiastically welcome their staff back in functional roles. If they don't, it may signal that they did not assign their best and brightest to project teams in the beginning. Consider this a warning sign that the product creation capability itself may need careful re-examination.

Nurture Tribal Knowledge

✳ ✳ ✳

Tribal knowledge captures the detailed "art" resident in the experts and specialists at many levels of an organization.

These individuals represent repositories of vast and valuable intellectual capital. For example, a tooling specialist may possess technical knowledge about the specifications that yield a particular finish, something that can be demonstrated by example but has not necessarily been specified in writing. Another individual may have particular insight unique to the company or industry, such as how to structure focus groups to achieve the best results for a particular product category like toddler toys. Such knowledge, though sometimes widely held, rarely gets documented. Often, those who have the knowledge do not have a full understanding of just how valuable their experience and expertise may be. Even if they do, few companies provide a formal mechanism for documenting this tacit knowledge. Every company should ensure that tribal knowledge is not lost when employees change jobs, leave the company, or simply retire.

Although implicit, such knowledge can be made explicit through a variety of techniques. Toyota maintains a "design guide" that not only provides specifications for new designs but also highlights design failures from the past. Pushing the edge of the envelope means breaking the seams occasionally. Process mapping of routine, but informal processes can often uncover steps that might not appear in the formal policy manuals. Pairing old sages with younger staff in an apprenticeship model also offers a way to transmit tribal knowledge through its most traditional route, word-of-mouth. Better yet, encouraging senior staff to keep journals and to share them with their apprentices keeps tribal knowledge alive.

Companies can also formalize knowledge capture through reviews, annually at a minimum and especially at the conclusion of significant projects. During these reviews, individuals should pause and ask "What have we learned about what we do, what works, and what doesn't? They should then record their answers in a group journal or database. This documented tribal knowledge should then be used to inform process design and training programs so that lessons learned contribute materially to continuous improvement.

Although tribal knowledge fosters continuous improvement, tribal "lore" can inhibit it. A naysayer may recount past failures as a rationale for resisting change, when the real resistance comes from the "not invented here" syndrome. Senior management should ensure that processes and decisions are grounded in valid knowledge and not just an old habit that no one questions any more. Like they do at Toyota, when senior managers ask "Why?" five times, they get five good answers when those responses come from knowledge and not lore. If the answers include a variation of "because that's the way we've always done it," it signals a reliance on lore, not knowledge.

Practice Project Management Discipline

* * *

While the soft skills of inspirational leadership foster an appropriately supportive culture, the hard skills of project management also play an important role in corporate product creation. Without the discipline of project management the organization will not deliver optimal business results. Without a nurturing environment, however, project management can easily yield a bureaucracy that cripples the innovative spirit. As true of most managerial challenges, success comes from finding the right balance between competing tensions.

Scaling the project management approach to the scope and complexity of the project is fundamental to effective project management. Though no two projects have the same characteristics, projects should be grouped into logical segments that explicitly identify the desired level of project management discipline. Figure 6.1 offers an example of some project classification guidelines to ensure the appropriate project management rigor. In this example, two factors drive a need for greater project discipline:

	CLASSIFICATION FACTORS		PROJECT MANAGEMENT PRACTICES		
TYPE	SCOPE	RISK	LEADERSHIP	STAFFING	REPORTING
MAJOR	• New platform • Major technology advancement • Staffing 25 or more fulltime equivalents	• New market/product for company • Unproven technology • Major capital and expense	Fulltime project leader with significant prior leadership experience	3 to 1 ratio of fulltime to part-time staff	Monthly written reports to executive leadership team and regular communication with the lead executive
REGULAR	• Block upgrade • Not a major technology advance • Staffing 10 to 15 fulltime and part timers	• Facing known competitors • Proven technology • 10% to 15% of total development budget	Fulltime project leader with prior experience as fulltime project team member	1 to 2 ratio of fulltime to part time staff	Monthly written status reports to lead executive with intermittent informal communication
MINOR	• Feature upgrade • Not a big technology advance • Involvement of 10 or fewer team members	• Facing known competitors • Proven technology • Less than 5% of total development budget	Part-time project leader with prior experience on project teams	All part-time staff	Informal reporting to the project manager

Figure 6.1 Project Segmentation and Management Practices

project scope and business risk. The project segmentation, in turn, provides guidance on three critical managerial practices: leadership, staffing mix, reporting process. Although this example, drawn from experience, will not apply across all companies and project types, it does provide guidance for tailoring a company's unique policy for project management discipline.

Recognize when a project becomes a project. Applying the appropriate project management discipline requires a comprehensive view of the full set of projects. This, however, raises a fundamental project management issue. When does a project become a project?

The question may sound academic and of little practical relevance, but experienced executives know that failing to answer the question wastes valuable resources and time in many organizations. All organizations have a host of activities underway at any given time, most not operating under the formal discipline of project management. These unacknowledged projects generally fall into the minor project classification depicted in Figure 6.1. As such, they warrant limited project management discipline. But *limited* project management discipline is quite different from *no* project management discipline.

In some instances, significant projects can proceed informally far longer than prudent. Senior managers must regularly examine the allocation of resources to identify unacknowledged projects that may require more management discipline. Some of these could include "pet" projects that draw on the discretionary time of a few staff and deserve to be nurtured. If, however, the project consumes the discretionary time of many individuals, the company's investment may have become significant enough to warrant a higher level of project management discipline. Such discipline, however, should not crush these unacknowledged projects. In fact, if an unacknowledged project has attracted the attention and time of a large number of talented staff, it could very well signal a potentially rewarding opportunity to pursue aggressively. Project management discipline's purpose is to bring the necessary focus to the project to ensure the company captures the opportunity, not impose bureaucratic overhead that stifles creativity.

Other projects may offer a major opportunity but be inconsistent with the business goals. Management should redirect these projects to support the business strategy, or consider licensing or spinning them off to an organization with a compatible strategy. Still other nascent projects may not require significant resources, just time to develop to the point where they can be assessed relative to strategy.

Recognizing the personal motivations that often underlie unacknowledged projects, senior management must exercise discipline with a deliberate but not heavy hand. Iron-fisted attempts to crush a project or to force unnecessary discipline can drive the project "underground" and out of management's sight. That benefits no one. On the other hand, if an unacknowledged project is surfaced and isn't blessed or redirected, the organization will assume that such activity is permissible and people will continue to bend the rules. Experience and good judgment are the keys to finding the appropriate balance between freedom and discipline in project management.

Most projects, of course, evolve from formal processes such as an appropriations request or a budgeting commitment. Product managers initiate minor projects for feature upgrades or special customer programs on a continuous basis determined by the needs of the competitive product portfolio. New platform designs generally evolve out of a major product strategy review. Significant block upgrades usually gain approval through the annual budget planning process. Although the level of project management discipline varies, all project leaders must have a clear understanding of how their projects will create value for the customer and the business.

Choose project sponsors and leaders wisely. Identifying a project sponsor and leader starts the project formalization process. The sponsor serves as the single focal point for decision making, sets parameters within which the project leader has authority to act, sets vision, clears obstacles, and supports the team. A product manager or another key executive often takes on the project sponsor role. For major projects, the sponsor could and probably should be at the executive vice president or business unit leader level.

Project sponsors have responsibility for selecting the right project leaders and coaching them throughout the process. Although the project leader has ultimate line responsibility for project success, the sponsor shares that responsibility from an executive perspective. Picking the best leader for the circumstances and providing the needed support, therefore, are critical project sponsor tasks. The project leader must have the ability to manage upwards, collaborate laterally, and lead the project staff. Specifically, the project leader has a broad range of responsibilities:

(1) Ensure achievement of project goals and objectives

(2) Staff the team with the proper mix of skills, knowledge, and experience

(3) Remove obstacles and garner needed resources

(4) Maintain project management discipline

(5) Provide guidance to the team in a timely and straightforward manner

(6) Conduct performance assessments of sub-team leaders and provide feedback

(7) Communicate project status continuously to senior management and the team

In many companies the leadership of large projects offers the path to executive management. Such projects provide excellent training for future company leaders since they involve integrating many functions, dealing with considerable ambiguity, meeting goals and objectives, and employing leadership skills and judgment. Few assignments offer a better training ground for general management.

Formalize the project charter. The next step for formalizing a project involves the creation of a project charter by the project sponsor and project leader. The project charter describes the objectives, resource requirements, timing commitments, and key participants. As shown in Figure 6.2, an example for the hypothetical Omega Corporation, the project sponsor and the project leader should clearly articulate the envisioned output of the project, the target market, and key business metrics like retail price point, cost target, projected volume, required capital and expense,

European Built-In　　　　*September 15, 2006*
Product Creation Project Charter

Project Description	Develop new Built-In Platform for the European market to replace the original platform used in both U.S. and Europe. Key new functionality includes voice-activated controls.		
Target Market	Professional men seeking a high-end unit for home entertainment. The product will be tailored to the unique European market seeking a high-tech design and compact footprint. To be sold through big box electronics retailers and independent specialty shops.		

Retail Price Point	$2,350	Development Budget	$14,800,000
Target Production Cost:	$1,650	Capital Budget	$23,500,000
Annual Volume Target	35,000 units	Return on Investment	52%

Department	Functional Lead	Resource Requirement	
Research	Isabella Rodriguez	30 person months	2 Fulltime
Engineering	Martin Hunt	190 person months	6 Fulltime
Marketing	Linda Insight	44 person months	3 Fulltime
Operations	Takeru Ohmae	52 person months	4 Fulltime
Procurement	Stefan Laub	24 person months	1 Fulltime
Finance	Taro Okebano	5 person months	No Fulltime
Customer Support	Sandeep Singh	5 person months	No Fulltime

Major Milestones	Target Date	Critical Planning Parameters
Idea Selection	October 1, 2006	ROI of at least 30%
Capital Funding	November 15, 2006	25% footprint reduction target
Concept Selection	January 1, 2007	Employ modular design for future upgradeability
Launch Date	May 1, 2007	Introduce hands free control technology
Full Ramp-Up	November 1, 2007	Design for recycling/sustainability
		New platform for later model expansion
		Low cost country manufacturing

PROJECT PHASE	2006	2007
	J F M A M J J A S O N D	J F M A M J J A S O N D
Ideation		
Concept Development		
Design and Build		
Launch		

Executive Sponsor	Project Sponsor	Project Leader
Francoise Dunn	Jim Wright	Rolf Wittenberg
President	Vice President	Development Director
Europe Region	Built-In Development Group	Built-In Development Munich

Figure 6.2 Sample Project Charter

return on investment, and critical planning parameters. In doing so, both commit to deliver the business results.

Next, the project leader and sponsor should identify key functional leaders to support the project and engage those individuals

151

to develop a resource plan, project budget, and timelines. Once completed, the charter should be approved by an executive sponsor who commits to help make the project a success by ensuring the necessary organizational support. As shown in the figure, a charter need not be detailed, but it should provide clear direction and reflect explicit commitments for deliverables and timing.

The project charter should contain a set of key metrics for managing the project and judging the end result. Metrics should include clear information on project timing and scope. For some projects, for example, the due dates are flexible because the required timing may depend on many other factors external to the project such as alternative product launch priorities, advertising cycles, or sales channel needs. Due dates may be the dominant consideration in other projects because of fixed scheduling demands or competitive action. Where timing is critical to success, the charter should allow some flexibility in project scope. If, however, the scope is very broad, the charter should allow some flexibility in the timing. Broad scope and rigid time constraints can be a recipe for failure. Whether the focus is on scope or time, the charter should set specific, clear goals supported by metrics tied to the team's annual performance review.

A good charter requires a delicate balance between defining sufficient specificity to focus the team on a clear objective and constraining the natural evolution of the project with too much detail. Poor charters operate at either extreme. At one end would be a vague mission statement such as, "to enhance competitive advantage," at the other end too many objectives rather than the "critical few." Other common errors include trying to do too much within the scope of a single project or setting overly aggressive cost and time estimates.

Some charters are inadequate because they reflect a narrow functional point of view such as an operations cost reduction or an engineering technical breakthrough rather than an appropriately cross-functional, general management view. Unless the company has a pervasive culture of cross-functional collaboration, project charters tend to fail because they do not involve all principal parties. The most egregious charter mistake, however, is initiating a product creation project that requires new invention rather than separating the invention from development.

Define the staffing and organization plan. Once the project charter has been finalized, the project sponsor and the project leader must define the organizational model and staffing plan and line up all necessary resources and commitments before formally launching the project. No project can succeed without strong functional support to provide the financial and staff resources needed by the project team. Executive leadership can help by encouraging a corporate culture that ensures that functional groups understand that they exist to serve the company's projects. With a supportive culture, the sponsor and leader can define an appropriate reporting structure within the company including matrix links to various functional groups.

They should also define an internal project structure that can accomplish the project goals and lend experience and support to key team members. With the help of a designated lead executive, typically a business unit leader or a senior functional executive, the project sponsor and leader should also confirm the necessary approval matrix to gain all needed functional endorsement at each stage of the project.

With an agreed structure and resource plan, the project leader can turn full attention to recruiting a team with the appropriate abilities. In addition to technical skills and business experience, the project leader should seek personal characteristics that show strong team building and collaborative skills. The team leader should personally screen potential core team members, much like the screening of new hires. In addition to personal interviews, discussions with past supervisors, input from peers, and personnel records provide useful insights for determining whether the individual will make a good fit. When considering fit, different projects demand different types of individual characteristics. Product creation projects have an execution focus and accordingly need results-driven people while research projects may need a greater emphasis on creativity.

Team members with experience in multiple disciplines are particularly valuable since they can handle a diversity of tasks if required to do so. This speeds up the entire project by easing the handoffs among members of a larger team composed of more narrowly focused functional experts. Individuals with a broader range of capabilities also offer a hedge against the loss of a team member—a practical consideration even in companies that have policies against reassigning team members during a project. Things happen: someone gets sick, a team member quits, or a key player gets a chance for a long-sought promotion.

As described in Figure 6.1, different size projects warrant a different mix of full-time and part-time staff. Full-time staff members bring a greater degree of focus and sense of commitment because they don't have competing demands, and their personal success depends solely on the project at hand. Projects with narrower scope, however, may not require a dedicated full-time staff member from a particular discipline or area. Full-time staffing in such an instance could result in the team member being under-utilized, or worse, creating "busy work" that impedes productivity.

Another solution might be to assign an individual to more than one project if the scope of the projects permits. Harvard Business School professor Steven Wheelwright and Harvard Dean Kim Clark[2] citing their own and others' research note that typically the greatest individual productivity occurs when project team members are assigned to two projects simultaneously. With a single project efficiency can be lost due to underutilization because of varying workload demands. Splitting time on multiple projects helps smooth the workload. Working on too many projects, however, can create inefficiencies because of a lack of focus or an imbalance among competing priorities.

Individual efficiency is only one consideration. In some cases, speed may be far more important than staff utilization. In other instances, the individual's personal capabilities or talents make the difference. Some team members may not handle multitasking well and could end up under-delivering on all projects if spread too thinly. Ultimately, management judgment must find the staffing mix that achieves the most critical goals of the project.

Project work ebbs and flows and the staffing plan must anticipate the need for ramping up and down to meet resource bubbles in a way that works for the project as well as the individual team members. Despite continuously changing resource needs, staff continuity is critical to successful project execution. Continuity increases the opportunity for concurrent development and

---

[2] Wheelwright, Steven C. and Clark, Kim B., *Revolutionizing Product Development*, The Free Press, New York, 1992, page 91.

reduces the time lost because of changing team members. In addition to getting up to speed, new members inevitably feel the need to challenge what has been done before, which causes rework—sometimes warranted and sometimes not. To ensure continuity, major projects should have at least one core team member from each function dedicated to the project and at least one backup to provide continuity if the core team member leaves.

Most companies have a reasonably well defined process for developing their pool of talented staff, especially for those at the management level and above. This process, designed to staff the organization optimally, enables individuals to progress through the organization to reach their full potential. The process typically centers around business units and functions. Because it represents such a rich environment for nurturing executive leadership talent, product creation project team membership should be an inherent rung in the management ladder. This has a positive impact not only on staffing projects with the right people, but also on the development of those people as future leaders.

Use red teams to leverage experience. The concept of a "red team" originated in the U.S. military where war game simulations pitted the active forces against a group of experts in the role of adversaries who challenge conventional military doctrine, plans, and tactics. The concept migrated from the military to military contractors who modified it for product creation projects. In that context the red team is composed of experts, but they do not behave as adversaries.

Red team members draw upon their depth of experience to challenge the project teams' thinking in a non-threatening way. Much as they would in a military exercise, the product creation red team provides the opportunity for a "safe practice" review for the project team. Comprised of experienced executives from other business units and retired managers, the red team has no authority over the project team. Instead, it offers the type of rich experience—and challenging issues—that the team will face from the senior management at its formal tollgate reviews. (See Figure 6.3 for an Omega Corporation example.)

Red teams also offer another way to sustain tribal knowledge by engaging company retirees as consultants. By engaging executives from other divisions and regions, the red team creates linkages among business groups that might not otherwise exist, opening the door for discovering unanticipated business synergies or sharing best practices. A properly formed red team represents diverse experience ideally gained at high management levels in the key functions important to the project—engineering, operations, information technology, marketing, procurement, program management, or finance. Effective red team members must have credibility with the project team and respect from the lead executive. Otherwise, their advice will be discounted or even ignored.

Red teams should not serve simply as an informal sounding board on an ad hoc basis. Instead they should have clearly

EUROPEAN BUILT-IN PROJECT RED TEAM	
Red Team Membership	**Meeting Dates/Locations**
John Dempsey, VP Corporate Procurement Vivek Patel, Retired VP of Engineering Gabriella Rodochonoci, VP Marketing, U.S. Taichi Uchida, Manufacturing Director, China Marco Castellano, Finance Director, Europe	August 15, 2006: Paris November 14, 2006: Milan March 6, 2007: TBD

Figure 6.3 Sample Red Team Plan

defined purpose and process including specified frequency and timing of interaction with the project team. Such interactions should be scheduled in advance of major project milestones and tollgates. Scheduling should allow sufficient time for the project team to address critical issues that the red team may raise. The red team only provides value if it helps the project team identify and resolve issues that could affect the ultimate fulfillment of its project charter.

To deliver maximum value, red teams should follow clear rules of engagement:

- Review the project to ensure it will meet the business objectives

- Foster an environment of open debate and no hidden agendas

- Consider all sides of an issue with objectivity

- Respect the confidentiality of information revealed during the red team review

- Conduct reviews in a timely manner to support critical decisions of the project

- Produce a brief report for the project leader and sponsor that captures key recommendations (usually four to six)

Even with a long history in a variety of contexts, red teams remain underutilized despite their demonstrated value. Resistance to their use from project managers could explain the shortfall. Project managers that are uninitiated to the practice typically complain that such reviews would get in the way of real work. On the other hand, those who have benefited from a well structured red team process become converts: a non-threatening, constructive review by true experts demonstrates its value consistently. Good red teams do not micro-manage the project team. Instead, they serve as an effective management tool for the project manager as well as the company.

Done well, a red team review provides several key benefits. It:

- Leverages "outsiders" to ensure a broader perspective on the project

- Engages experienced resources as support for inexperienced staff

- Transmits tribal knowledge that may have gone undocumented

- Aids decision-making by challenging status quo thinking

- Mitigates risks by identifying and addressing critical issues

- Helps avoid surprises for the project team and management during the stage-gate process

In short, the red team concept offers a wealth of benefits with very low risk and minimal implementation effort. Because no other project management tool offers such high return for such a low investment, companies that do not use red teams for all major projects do themselves a great disservice.

Choose the right project support tools. Although project managers might have personal favorites among the host of possible project management tools, companies should insist on standardization. This is one place where creativity and independence needs constraint. Project leaders have many software tools available to support the management, documentation, collaboration, and communication. Using off-the-shelf software rather than custom designing home-grown systems nearly always proves more efficient and less costly. The systems that support the product creation process rarely create a competitive advantage; using the tools more effectively than others can. In addition, the use of software that conforms to industry standards facilitates collaboration with outside partners.

Project teams should also use a standard format for reporting project progress to enable senior management to quickly find and easily understand all of the key relevant information. The report should parallel the project charter, including project goals and key performance metrics. Figure 6.4 provides an example of a standardized project status report for the hypothetical Omega Corporation's product creation project shown in the charter example of Figure 6.2. It repeats the goals from the charter and adds status information on critical planning parameters, resources, and milestones. Back-up pages can provide additional details, but a single page "snapshot" report keeps senior management informed in a simple and consistent manner.

Status: Green **European Built-In** *November 1, 2006*

Product Creation Project Status Report

Project Description	Develop new Built-In Platform for the European market to replace the original platform used in both U.S. and Europe. Key new functionality includes voice-activated controls.			
Target Market	Professional men seeking a high-end unit for home entertainment. The product will be tailored to the unique European market seeking a high-tech design and compact footprint. To be sold through big box electronics retailers and independent specialty shops.			

	Original	Updated		Original	Updated
Retail Price Point	$2,350	$2,150	Development Budget	$14.8 M	$14.9 M
Target Cost:	$1,650	$1,450	Capital Budget	$23.5 M	$24.8 M
Annual Volume Target	35,000	35,000	Return on Investment	52%	49%

Department	Total Requirement	Planned To Date	Actual To Date
Research	30 person months	9 person months	7 person months
Engineering	190 person months	40 person months	38 person months
Marketing	44 person months	14 person months	12 person months
Operations	52 person months	5 person months	7 person months
Procurement	24 person months	6 person months	7 person months
Finance	5 person months	1 person month	1 person month
Customer Support	5 person months	1 person month	1 person month

Major Milestones	Target Date	Current Target Dates	Comments
Idea Selection	October 1, 2006	November 1, 2006	Slow staff ramp-up
Capital Funding	November 15, 2006	December 15, 2006	Due date shifted
Concept Selection	January 1, 2007	February 1, 2007	Due date shifted
Launch Date	May 1, 2007	June 1, 2007	Due date shifted
Full Ramp-Up	November 1, 2007	November 1, 2007	Production plan holds

Critical Planning Parameters	Status and Issues
ROI of at least 30%	ROI estimate of 49% for 4 year life (38% 3 year life)
25% footprint reduction target	33% footprint reduction expected
Employ modular design for future upgradeability	Four major modules envisioned in current design
Introduce hands-free control technology	hands-free control technology now on ADT
Design for recycling/sustainability	Expected to be 80% recyclable
New platform for later model expansion	Good/better/best price points being developed
Low cost country manufacturing plan	Examining sites in Romania and Poland

	2006	2007
PROJECT PHASE	J A S O N D	J F M A M J J A S O N D
Ideation		
Concept Development		
Design and Build		
Launch		

KEY ISSUES	MITIGATION PLANS
Delay in Ideation stage due to staffing shortfall	Staff on board; faster ramp avoids overall delay
Market price dropped from $2,350 to $2,150	Lower cost target by $200 so same profit per unit

Figure 6.4 Sample Project Status Report

Train for Success

* * *

Good process discipline demands good training. The best organizations go beyond technical knowledge or even custom training on the company's product creation stage-gate process. They also use training in leadership, project management tools, and communication skills to support successful project management. Every project leader—and potential future project leaders as well—should complete a comprehensive training program focused on these much needed skills.

Functional team leaders assigned to full-time roles should also consider the same training curriculum. Because most of these skills have value that extends far beyond the product creation process, the training can be part of a general corporate program. Experience suggests that the best training programs engage the participants in interactive discussions and exercises and also explore real-life company case examples—both good and bad.

Part-time and temporary functional team members and supplier representatives usually need only abbreviated training programs in the product creation process and project skills. Some level of training should be mandatory, however, to ensure that all project team members have a common language and understanding of the process and tools used. Senior management should also complete condensed versions of such training and would ideally participate in the training of others. Such involvement in training reinforces the importance of product creation and good project management to all participants.

Good training is not an ad hoc event. It results from a well-defined curriculum used throughout the organization. Full-length and abbreviated training modules can be "mixed and matched" to create customized training while ensuring consistency of concepts and methodologies company-wide.

Extending the Logic

* ❄ *

As highlighted many times in this and prior chapters, effective product creation requires an external view and often demands collaboration with outsiders. Although all of the concepts discussed so far apply to internal as well as external efforts, the following chapter examines the unique concerns of innovation partnerships to bring focus to this critical opportunity.

Chapter 7

INNOVATION PARTNERSHIPS

————— ✳ ✳ ✳ —————

Connecting the Extended Enterprise

Even the best innovators should not go it alone. Under the leadership of Chief Executive Officer A.G. Lafley, the traditionally inward-focused Procter & Gamble® now seeks to derive half of its invention from external sources. Says Lafley, "Inventors are evenly distributed in the population, and we're as likely to find invention in a garage as in our labs."[1] P&G has employed a range of relationship models to develop innovative new products including the SpinBrush™ battery operated disposable toothbrush and Glad® Press 'N Seal™ food wrap.

The SpinBrush technology traces its roots to a spinning lollipop developed by inventor-entrepreneur John Osher. After some initial successful trials of the toothbrush—not the lollipop—at a couple of major retailers, P&G acquired the entrepreneur's

---

[1] Sellers, Patricia, "P&G: Teaching an Old Dog New Tricks," *Fortune*, May 31, 2004.

company, Dr. Johns Products, Ltd. With P&G's added marketing and distribution muscle, the toothbrush became the number one brand in the United States.[2] Following a completely different path, P&G chose to exploit its internally developed adhesive technology—used in packaging and Crest® Whitestrips®—by partnering with a competitor. An auction of the patented P&G technology in 2002 resulted in a joint venture with Clorox®—the owner of the Glad brand—to produce Glad Press 'N Seal food wrap which has displaced S.C. Johnson's Saran™ wrap as the U.S. market leader.[3]

Throughout its history, the automotive industry has relied on suppliers to develop new technologies—from headlights in the early 1900s to anti-lock brakes and cruise control in the 1960s to heads-up displays today.[4] Consumer electronics companies like Apple and Hewlett-Packard have relied on industrial designers such as IDEO and Smart Design for decades. IDEO created the first production mouse for the Apple Macintosh® in the 1980s.[5] More recently, Smart Design helped HP design the ScanJet 4670, which sits in an easel holder but can scan an oversized book or a picture hanging on a wall.[6]

Successful innovation partnerships avoid the thorny challenges of excessive dependence on an outsider or the unintentional creation of a new competitor. A poorly conceived or poorly managed partnership can lead to disastrous results. The

---

[2] http://www.betterproductdesign.net.

[3] Sellers, Patricia, "P&G: Teaching an Old Dog New Tricks," *Fortune*, May 31, 2004.

[4] http://www.motorera.com; http://inventors.about.com.

[5] http://www.ideo.com.

[6] Nussbaum, Bruce, "Annual Design Awards," *BusinessWeek*, July 5, 2004.

story of how IBM outsourced the personal computer operating system to Microsoft has been retold many times, although often with the unique clarity afforded only by 20/20 hindsight. A less familiar, but equally telling story begins with the intense competitive pressure placed on U.S. television manufacturers by Asian producers in the 1980s and 1990s. Zenith®, a leading producer of color televisions at the time, struggled throughout the 90s and ultimately sought Chapter 11 bankruptcy protection in 1999. It emerged later that year as a wholly owned subsidiary of its major supplier, the Korean *chaebol* LG Electronics, now a global producer of branded consumer electronics. Ironically, LG's relationship with Zenith dates back to the days when LG— a $17 billion global consumer electronics leader in 2005—was better known as Lucky Goldstar, a low-end, low-priced private label producer.

Although key strategic relationships capture the greatest visibility, many of the most successful innovation efforts derive from more mundane uses of a company's regular suppliers. Effective execution dominates in this context because any advantages gained can be short-lived if competitors, as they often do, incorporate the innovation in their own next design cycle. Nonetheless, a company that fails to exploit the innovation capabilities of its suppliers effectively can easily fall behind in the competitive race. As the Red Queen explained to Alice during her visit to Wonderland, "Now, here, you see, it takes all the running you can do, to keep in the same place."[7]

---

[7] Lewis Carroll, *Alice in Wonderland*, Gramercy Publishing, September 7, 2004.

Select Partners for Strategic and Cultural Fit

* ※ *

The horror stories of failed alliances may inhibit some from pursuing innovation with outsiders. Although a rigorous strategic approach to selecting partners does not guarantee success, it certainly improves the odds. As with any new product creation effort, the process must start with customer needs: What problems are we trying to solve? From there the process shifts to searching for existing technologies that could meet the customer needs, or could be modified to do so.

Craig Wynett, General Manager for Future Growth Initiatives at Procter & Gamble argues that most problems have already been solved by someone, somewhere. "At P&G, we think of creativity not as a mysterious gift of the talented few, but as an everyday task of making non-obvious connections, bringing together things that don't normally go together."[8] Accordingly, finding and using existing technology can be just as important as creating new technology. Wynett encourages his innovation teams to conduct rigorous searches for analogous problems and the resulting solutions. Reviewing patent filings can uncover a wealth of information but can also overwhelm the researcher. Broader searches of published media for innovations in other industries can help narrow the task or even identify new solution opportunities.

More typically, companies simply turn to existing suppliers for a technical solution to support an innovative concept. A

---

[8] http://www.smartlink.net.au.

long-term relationship with a supplier, however, does not necessarily create an adequate basis for an innovation partnership. A company's sourcing strategy should indicate clearly—and in advance of the specific need—whether or not a specific supplier should be involved in technology development. For example, a supplier may bring a cost leadership position to the table, but not have the appropriate capabilities to support an innovation partnership. In such a case, encouraging that supplier to assume an innovation partnership role could prompt an investment in resources that might compromise the supplier's cost leadership position and probably not prove particularly productive for either company.

Toyota's relationship with the component supplier ©Denso offers an excellent example of a long-standing innovation partnership that works. In 1949, Denso (formerly Nippondenso) was spun off from Toyota because of the component manufacturer's financial troubles. In 2005, however, the $26 billion company ranked as the third largest automotive supplier in the world (behind Bosch® and financially troubled Delphi) and depended on Toyota for less than half of its global revenues.

Despite the diminished dependence on Toyota, which owns only 24 percent of Denso, the supplier continues to deliver an innovation advantage for Toyota. For example, when Denso developed its global product line of alternators in the 1990s, its engineers began with intense discussion with Toyota regarding long-term, global vehicle plans. Based on its unique insight into Toyota's needs, Denso built a family of alternators using a single design concept but with a range of options—including three housings, nine wiring specs, four regulators, and 30 terminals—that would cover the needs of all Toyota vehicles. Leveraging this basic design concept, Denso offers 700 different alternators for sale to a wide variety of customers. The original design concept, however, was clearly developed to address Toyota's specific needs.[9]

Consider the softer side issues. In addition to developing a clear strategic mindset and a rigorous sourcing process, a company should also consider the softer side issues before investing in an innovation partnership. Although two companies may offer complementary innovation assets, a poor fit culturally could derail efforts for effective collaboration. For example, companies may have different time horizons for innovation: one focused on immediate market impact and the other on long-term technology breakthroughs. Such differences can yield mutual benefits, but only if these differences are recognized from the outset and the objectives of the partnership clearly plot the path the partners will follow.

Companies with distinctly different objectives can find mutual benefit if the innovation project achieves both sets of objectives without significantly diminishing either. For example, Whirlpool and Proctor & Gamble co-developed the technology for a product that refreshes clothes that would otherwise require dry cleaning. Whirlpool's objective was to sell and install new equipment, while P&G sought to benefit by supplying the installed product base with the cleaning products the equipment required.

Other differences—such as a legalistic versus a "handshake" approach to partnership or a competitive versus team-oriented

---

[9] Kamath, R.R., and Liker, J,K., "A Second Look at Japanese Product Development," *Harvard Business Review*, November-December 1994.

approach—offer little opportunity for synergy but can present major hurdles to successful collaboration. At one extreme, differences in ethical standards between two companies can present an insurmountable obstacle, but even small, tactical issues can impede effective collaboration. For example, one company that needed a creative, free-form approach to a particular issue struggled to integrate a supplier that preferred to work in a highly disciplined, structured way.

Ultimately, to succeed, an innovation partnership must offer benefits for each party. These should be articulated in writing at the concept stage of each collaborative project to avoid the kind of misunderstandings that inevitably arise from ambiguous relationships. Such clarity typically demands a senior executive's strategic perspective: project teams, in the throes of meeting near-term milestones, can often lose the bigger picture perspective that mandated the need for the relationship in the first place. An effective innovation partnership must result in a win for both partners.

Use Market and Technology Plans to Drive Relationships

* * *

The most productive innovation partnerships involve long-term, ongoing collaboration. Even the previously cited P&G and Clorox venture, which was limited to collaboration around a single adhesive technology, is expected to produce a series of products beyond the Glad Press 'N Seal product. Technology roadmaps, as discussed in Chapter 3, help ensure that the development efforts of the joint venture fully exploit a new technology, but in ways that complement the product and market plans of both companies.

Technology roadmaps prove equally critical to more traditional customer-supplier relationships. Ford, for example, uses technology roadmaps of safety devices to guide the innovation processes at suppliers of restraint systems. The roadmap indicates the planned rollout of new features such as side-air bags and window curtains so that the supplier focuses creation of new products in the right order and on the needs of specific automobile models. Designing a new side airbag for the Mustang would be a waste if Ford has made a window curtain in the Expedition sports utility vehicle its main priority.

Whirlpool, like many companies, has long offered supplier training in quality assurance to ensure robust designs. More recently, however, Whirlpool began offering supplier training in its internal innovation process including a heavy emphasis on ensuring consistency with "brand image." Suppliers now understand whether a new idea best fits the consumer positioning of the core Whirlpool brand or if it applies more appropriately to the higher-end KitchenAid brand. To lead this effort, Whirlpool created a small supplier innovation group, headed by a director level manager, not only to evaluate suppliers against an aggressive set of innovation goals, but also to teach them how to reach those goals.

Motorola and Black & Decker® sponsor technology fairs for their suppliers to facilitate two-way sharing of innovations.[10]

---

[10] Laseter, Timothy M., *Balanced Sourcing: Cooperation and Competition in Supplier Relationships*, Jossey-Bass, 1998.

Suppliers staff one hundred or more booths where they present new ideas that the engineering community might find of interest. Black & Decker calls its event, which includes its top 100 suppliers, "Product Awareness Day." Motorola has hosted as many as 180 suppliers in a huge tent erected on the parking lot of its corporate campus.

Balance the Mix of Top-Down Targets and Design Competitions

* ✳ *

Companies use different techniques for leveraging the innovation capabilities of their supply chain partners. Some specify design requirements up front. Others articulate an end goal but stop short of a set of hard specifications. A mix of both approaches can often yield the best outcomes over the long term.

Although market and technology plans help guide the innovation partnering process, collaboration also benefits greatly when companies set very specific design goals up front. In the automotive industry, for example, design goals often set new standards for cost, weight, and performance and can involve multiple suppliers. On the flip side, one automotive component manufacturer set out to develop a new universally applicable windshield wiper motor at a cost reduction of 30 percent by involving the suppliers of die-cast housings, electronics, and plastic parts. Rather than specifying individual targets for each, the company encouraged all of the suppliers to think beyond their traditional scope of experience and expertise. The approach led to creative solutions including the integration of previously separate parts to meet the overall system goals without squeezing supplier margins (the more traditional, albeit unfortunate, method of cost reduction in the automotive industry).

Fully outsourced innovation projects, such as Black & Decker's development of a branded glue gun in the late 1990s, require special attention. Black & Decker began with consumer research to create a precise set of functional requirements, including a price target. The supplier partner, a leading glue gun manufacturer, was then charged with developing and manufacturing the product to specification and consistent with Black & Decker's style and branding. In similar situations, when developing a new version of an existing product, companies also often employ benchmarking of competitive offerings for added market insight.

Financial targets. After defining the appropriate functional specifications, companies typically turn to financial targets, which cover three dimensions: cost, price, or value.[11] Price-based targets use price comparisons of competitive products, potentially with adjustments to reflect different features. Retailers often use this technique when introducing private-labeled versions of existing consumer products.

Cost-based targets work "bottoms-up" using the cost of the individual components to develop an appropriate cost for the end product. Used extensively in the automotive industry and in contract electronics manufacturing, this technique ensures that

---

[11] Ibid.

suppliers do not extract excessive margins from the customer. It may, however, inhibit innovative thinking. On the other hand, for some consumer goods, the bottoms-up cost model would shock many retailers who focus on price-point rather than cost.

Finally, value-based targets try to avoid the narrow thinking that can result from focusing too closely on existing product designs rather than on the value to the consumer. For example, Swatch S.A.® used a value-based approach to drive the design of its original offering, investing in those product components that delivered significant value to the consumer while minimizing the cost of those that did not.

Very structured processes linked to the tools for quality function deployment (QFD) can help make the translation from customer needs to component costs.[12] QFD employs a "house of quality" matrix that starts with consumer values such as "light weight" and "portable operation." The process then examines how each component contributes to each consumer value. For example, in a personal computer, a larger battery would improve "portable operation" by extending the amount of time that the PC would run off of the battery. The larger battery, on the other hand, would negatively affect the goal of "light weight" thus forcing a tradeoff. Other components might be smaller and require less electricity, which positively affects both values. With an understanding of consumer priorities, which always include cost, the designers can allocate the cost priorities to the components that contribute most to value. Although detailed QFD analyses to

---

[12] For more information, see "Setting Supplier Cost Targets: Getting Beyond the Basics," by Laseter, Ramachandran, and Voigt, *strategy+business*, Q1 1997.

support value-based target setting may not be warranted in all cases, it does provide rigor to an otherwise subjective process. More important, effective target setting matches the appropriate technique to the task at hand: each technique can be employed with rigor or more philosophically. The characteristics of the end product, the sourced component, and the overall project goals define the approach with the greatest benefit.

Design competitions. As an alternative to functional specifications and economic targets, companies can also use design competitions to encourage creative thinking while avoiding overdependence on a single supplier. For its major development projects, for example, the U.S. Department of Defense (DoD) holds competitions among a pre-selected set of two or three major suppliers. The government pays for each competing team to develop a design, which the government then owns. DoD can synthesize the best aspects of all designs and source the actual production to the most capable supplier. Often, however, companies will fund their own development projects in hopes of gaining some proprietary advantage.

The Joint Strike Fighter offers an excellent example of effective use of a design competition. In developing the aircraft, designed to serve both the United States Air Force and Navy, the DoD held a design competition among the leading engine manufacturers: Pratt-Whitney, General Electric, and Rolls-Royce. Pratt-Whitney earned the initial award for early stage development, but GE and Rolls Royce ultimately collaborated to develop a second engine configuration. The design competition kept all three core suppliers engaged for the long-term while ensuring a healthy level of competition.

Although a powerful technique for extracting multiple alternatives for a new technology, design competitions are rarely used outside of the defense industry. The long time horizons—major weapon systems can take well over a decade to develop and then stay in commission for decades afterwards—and the disproportionately powerful role of the U.S. government creates an environment distinctly different from other industries.

Nonetheless, the technique seems inappropriately underutilized given an increasingly knowledge-intensive economy, where innovative ideas can bubble-up from almost anywhere—large companies, small start-ups, and even individual inventor-entrepreneurs. Consider the environmental context of genetically-engineered drugs and the relationship between "big pharma" and the numerous biotech start-ups, both of which face similar long-term development efforts. Perhaps some funded design competitions would prove beneficial for all.

Manage Intellectual Property and Branding Carefully

* ** *

Intellectual property ownership is one of the most contentious issues in collaborative development projects. Theoretically, the issue can be resolved contractually for each individual innovation partnership, but a company's reputation built on a history of past transactions—in some instances transgressions—can greatly influence its available opportunities for collaboration.

At the transgressions end of the spectrum, executives of automotive industry suppliers openly railed against the practices of Ignatio Lopez during his reign as chief procurement officer at General Motors. Lopez proclaimed a new era in supplier relationships in the early 1990s by tearing up existing contracts and aggressively re-bidding the parts. Although he drove a billion dollars to the company's bottom line during his tenure, GM was ultimately ostracized by many of the industry's technology leaders. One supplier executive openly stated that his organization would no longer take its best ideas to General Motors for fear that its intellectual property would be handed over to another supplier that did not invest in technology development and could therefore produce the part more cheaply[13].

At the other extreme, the venture capital industry in Silicon Valley operates on a high degree of implied trust. Leading venture capitalists (VCs) refuse even to sign a nondisclosure agreement (NDA) to protect an entrepreneur's intellectual property. Because leading VCs see hundreds of new business ideas every week, inevitably some are similar. Signing an NDA in advance of hearing a pitch from an entrepreneur limits the venture firm's ability to defend itself when that supposedly new idea is already under development by someone else. Despite the risk to intellectual property, the system obviously works. Entrepreneurs quickly come to accept the conditions because the upside outweighs the downside. If, however, a venture capital firm builds a reputation for consistently appropriating innovative ideas, it surely would lose the innovation idea flow necessary for survival.

---

[13] Sorge, Marjorie, and David C. Smith. "Lopez mystique matches Garbo's. (General Motor Corp. VP for worldwide purchasing J. Ignacio Lopez de Arriortua)." *Wards Auto World* 28.n9 (Sept 1992):81(2).

In an attempt to increase the flow of new ideas from its existing supply base, Whirlpool has addressed the intellectual property ownership issue head-on. Suppliers said that the company's bias towards total intellectual property ownership from all joint development efforts inhibited their willingness to invest in new projects. Whirlpool softened its stance on the ownership issue and, more important, started communicating the company's new position throughout the supply base. Although individual projects may still be contracted under different terms when appropriate for both parties, the new policy encourages collaboration by explicitly stating expectations regarding intellectual property and how suppliers will recover their investments.

Whose brand is it? Branding presents a related issue in innovation partnerships. In addition to control of intellectual property rights, the partners must think strategically about branding. While "merely" a component supplier, Intel Corporation has established a highly visible brand position with consumers through the "Intel Inside" logo on personal computers. Although Intel offered price discounts in exchange for co-branding, the tactic has created a significant barrier to re-sourcing by the computer manufacturers. Intel commands an 80 percent share of the PC microprocessor market. Intel's branding advantage also helps explain why the company generated 22 percent net profit margins in 2005 while the computer industry as a whole (including Dell) averaged less than three percent[14].

As companies shift greater product creation responsibility to innovation partners, they must exercise thoughtful control of branding issues or risk a shifting of power in the relationship.

---

[14] http://premium.hoovers.com/subscribe/co/fin/comparison.xhtml?ID=fffryrsyrrcyrhrsh

Johnson Controls, a leading maker of automotive seating systems, conducts it own consumer research and considers cars simply as carrying vehicles for its products. To date first-tier automotive suppliers have not established a consumer brand image, yet such end-product research could ultimately lead to supply-base power like the strongly branded diesel engines of Cummins®, Detroit Diesel®, and Caterpillar in the truck industry.

In fact, branding can drive an innovation partnership as demonstrated by the collaboration of P&G and Clorox on Glad Press 'N Seal food wrap. Such efforts, however, require careful branding strategies. For example, consider Coca-Cola's® search for the right co-branding balance as it reformulated its Diet Coke® product line using new artificial sweetener options.

In the early 1980s, the Diet Coke formula incorporated a variety of sweeteners including saccharin, aspartame, and a blend of both. On product packaging and advertising, the company also vacillated between generic branding and co-branding with the NutraSweet® logo from G.D. Searle, the inventor of aspartame. Although Diet Coke ultimately dropped the NutraSweet logo, it faced the same predicament with the introduction of the artificial sweetener, sucralose trademarked as SPLENDA® and produced by McNeil Nutritionals, part of pharmaceutical and health care giant, Johnson & Johnson. New diet cola products introduced by Coca-Cola in 2005 included Coke Zero® sweetened with aspartame and acesulfame potassium without any co-branding, Diet Coke sweetened and co-branded with SPLENDA®, and "regular" Diet Coke with aspartame only and no longer co-branded with NutraSweet®[15].

---

[15] Anonymous, "Battle of the Diet Cola Clones." *Consumer Reports*, Volume 70, Issue 12, page 9. (December 2005)

While offering such variety may appear important to the success of the products and Coke's profits, it also can confuse consumers, a fact not lost on the competition. In an apparent attempt to imply that Coca-Cola is indecisive in its product offering, Pepsi® launched a television advertising campaign based on the song lyric "Did you ever have to make up your mind?"—a direct swipe at the several different products that resulted from Coke's sweetener dilemma.

Engage the Supplier According to Product Needs

* * *

When searching for a new technology to be incorporated into an existing product, a heads-up display in a car for example, companies should take into consideration the complexity of how the new technology interfaces with other aspects of the end product. Original automotive industry research provides a useful framework and taxonomy.[16] At one extreme the desired technology could be a *hidden component*, one that has a simple set of interfaces with the end product and is relatively invisible to the consumer—behind the "green line" in Whirlpool's terminology. Examples would include the gas cap in an automobile or the power cord for a major appliance (see Figure 7.1).

---

[16] Laseter, T., Ramdas, K. and Swerdlow, D. "The Supply Side of Design and Development," *strategy+business* (31): 20-25 (2003).

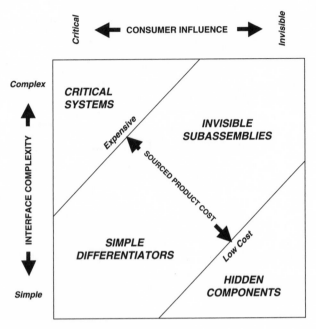

Figure 7.1 Product Element Classification

At the other extreme, the product element could have many critical interfaces with the end product and be a major influence on the consumer's perception and experience with it. These are called critical systems, and examples include an automobile's transmission, which defines many key performance characteristics, or a refrigerator's compressor, which affects noise and temperature control. In the case of complete design outsourcing, the entire product would be considered a *critical system*.

Other elements, such as a wiring harness in an automobile or appliance, may prove just as complex (and often just as expensive) as *critical systems*, but not have a great deal of consumer influence. These are called *invisible subassemblies*. Finally, while a host of simple elements, such as decorative treatments on an automobile dashboard or an appliance control panel may have fairly simple interfaces with the key functional parts of the end product, they can still exert significant influence on the consumer's perception or experience. These *simple differentiators* make up the fourth category of product elements.

This taxonomy helps clarify the appropriate approach for integrating the partner into the innovation process as shown in Figure 7.2. As expected, critical systems demand the most comprehensive design process integration, and the supplier/partner

	Critical Systems	Hidden Components	Simple Differentiators	Invisible Subassemblies
Design Process Integration	Comprehensive Integration	Limited Integration	Concept Design Focus	Financial Planning Focus
Involvement Timing	Early And Continuous	Later, After Key Areas Finalized	Early But Declining	Early With Lagged Iteration
Design Philosophy	Designed by Expert (Inside or Outside)	Catalog or Supplier Design	Concept Inside, Supplier Details	Black Box to Expert Suppliers
Specification Approach	Performance Specifications	Physical Specifications	Mixed Specifications	Mixed Specifications
Product Line Complexity	Few, possibly unique to end product	Single version for the end product	Many versions for each end design	Few and shared across multiple designs
Prototype Testing	Integrated at all stages	By supplier to customer specs and in concept development	Integrated to ensure aesthetics met	Fully delegated to supplier until product prototype testing

Source: Adapted from "Product Types and Supplier Roles in Product Development: An Exploratory Analysis" by Laseter and Ramdas, IEE Transactions on Engineering Management (49:2) (May 2002).

Figure 7.2 Summary of Distinct Partner Roles in Product Creation

should be integrated into the process very early and continuously. By contrast, simple differentiators require early involvement at the concept design and styling phase but then can be kept separate because the interface complexity is low.

Hidden components should be dealt with later in the innovation process once key design parameters stabilize. Invisible subassemblies typically require close financial monitoring because they can represent a significant proportion of the end-product cost but do not drive the key design decisions. In many cases, the design team should simply try to incorporate an existing version of an element into the new product rather than waste resources developing a new version that will have little impact on customer perception or experience. Rather than apply pressure on suppliers to make design changes, the development team might challenge them instead to propose potential cost reductions or quality improvements.

Critical systems can offer the greatest opportunity for an innovation alliance when a company can harness the capabilities of an existing expert as Hewlett-Packard has done in its alliance with Canon™ for printer "engines." In dealing with such experts, the engaging partner should not attempt to offer detailed guidance, such as drawings and specifications, but should focus instead on defining the appropriate performance parameters. In other words, clearly define the desired end result, but leave the details to the expert.

Simple differentiators also offer excellent opportunities for leveraging outside experts. For example, many of the ideas generated by an industrial design firm do not affect the "guts" of the product, but their aesthetic appeal can differentiate it in the marketplace. Crown Equipment Corporation used an outside industrial design firm, RichardsonSmith, to bring style and improved functionality to the staid fork-truck industry.[17] Thoughtful, but ultimately simple, adjustments such as changes to the shape and color of control knobs and the placement of lights helped the product stand out from the competition and earn a price premium. When Crown began making material handling equipment in the late 1950s, it had 150 employees and $1 million in sales. In 2005, by focusing on design differentiation, Crown was the leading producer of electric heavy-duty lift trucks with 6900 employees and more than $1 billion in sales.

The four product element categories (refer to Figure 7.1) also help clarify key design priorities such as component standardization and testing policies. For example, invisible subassemblies should be treated as modules and standardized for the most part and even used across multiple end-product designs where possible to control costs and simplify interface complexity. The supplier should maintain full control of the testing to enable a plug-and-play mindset in development.

Given their typically lower cost, hidden components can be "value engineered" to meet the specific needs of the end-product design and their suppliers held accountable to detailed test specifications set by the customer. Where warranted, to achieve differentiation, multiple versions may be appropriate for critical systems. Testing should be done at the end-product level given the element's criticality to the end-product application. A wide variety of simple differentiators can give the consumer a perception of choice at a relatively low cost. The key testing issues

[17] Freeze, Karen J., "Crown Equipment Corporation: Design Services Strategy," Design Management Institute Case Study, 1991.

include aesthetics and functionality rather than the element's integration into the product performance. In all cases, designs must be tested in the product design stage to assure that performance, integration, interoperability and brand image goals are met.

Also important, the classification of a product element does not remain static. It changes over time because of competitive dynamics and consumer perceptions. For example, electric automobile windows offered a basis for differentiation when first introduced but have now become "hidden" from a consumer point of view. Accordingly, the product creation role for the supplier of electric windows must also evolve.

Continuously Renew or Replace Innovation Partners

* * *

Regardless of the strength of existing partner relationships, companies should continuously scan the horizon for potential new innovation partners. During the height of its growth in the new economy, Cisco Systems acquired an average of about 20 companies per year, a major contributor to the company's phenomenal growth from a $2 billion company in 1995 to $22 billion six years later. These acquisitions played a critical role in the company's innovation strategy despite Cisco's expenditure of approximately $1 billion per year on research and development at the turn of the century.

In addition to the training Whirlpool provides to its full supply base, the company offers even richer insight to a set of core suppliers on its North American and European supplier councils. At tri-annual meetings of the Whirlpool North American Supplier Council, the company presents a broad range of business plans and updates to the suppliers covering everything from brand strategy to retail trends to quality and manufacturing issues. Meetings in corporate headquarters in Benton Harbor, Michigan, have afforded opportunities to demonstrate new products in the Whirlpool Development Center. Other meeting locations, such as at Whirlpool's fully-owned subsidiary in Mexico or its "Insperience™ Studio" featuring Whirlpool and KitchenAid products in fully-functioning kitchens and family studios in Atlanta, allow the company to focus the suppliers' attention on areas of particular interest as needed.

Membership on the supplier councils shifts over time as Whirlpool's supply base changes to fit the company's needs. Whirlpool has also formed a technology council, which includes a variety of suppliers—both current and potential future innovation partners. The key is to maintain an appropriate balance by maintaining a long-term view of innovation partnership while remaining open to new partnership opportunities.

In short, this principle simply closes the loop back to the original admonition to apply a strategic approach in selecting innovation partners. The continual emergence of new innovations and the inevitable evolution of priorities within any partnership, demand an ongoing search for new possibilities. Though less-well funded than at the peak of the Internet boom, many companies maintain venture capital units to seek out and partner with emerging companies.

Formed in the early 1990s, Intel Capital had in 2005 a portfolio of approximately 200 companies with an equity value of nearly $1 billion dollars in three separate funds. The largest, the $500 million Intel Communications Fund, focuses on "technology companies developing innovative networking and communications solutions." The quarter-billion dollar Intel® 64 Fund supports the company's Itanium®-based server and workstation business, while the $200 million Intel Digital Home Fund invests in companies that can help Intel drive convergence of personal computer and consumer electronics devices in the home.[18]

Worth the Effort

* * *

Innovation partnerships should employ the same key techniques required for internal product or service development, but across an extended enterprise. As such, the challenges are significantly greater and probably should not be pursued by a company without an already strong internal product creation discipline. Managing creative individuals within an organization is challenging enough; the added complexity of managing outside innovation resources could prove overwhelming.

---

[18] http://www.intelportfolio.com/cps/portlist_fund.asp.

Challenging to master, innovation partnerships warrant the effort. American author John Gardner said, "Mastery is not something that strikes in an instant, like a thunderbolt, but a gathering power that moves steadily through time, like weather."[19] Companies that invest the time and effort required to build and sustain innovation partnerships will help ensure that their product creation efforts flourish.

---

[19] *The Concise Columbia Dictionary of Quotations*, Robert Andrews, editor, Columbia University Press, New York, 1989.

Part II
CASE STUDIES

Chapter 8

WALT DISNEY IMAGINEERING

* ✳ *

Stories of the Impossible

The origin, evolution, and current state of Walt Disney Imagineering is a story worth the telling—and those words are not used lightly. Storytelling defines the essence of Imagineering—its very reason for being. That was Walt Disney's vision and Imagineering continues to live it more than half a century later.

Imagineer[1]

You won't find it in the dictionary. But any Imagineer can tell you the word is both a verb and a noun. To Imagineer. To be an Imagineer. Imagineering has become a purely Disney word. The name combines imagination with engineering to describe both what we do and who we are.

---

[1] *Walt Disney Imagineering: A Behind the Dreams Look at Making the Magic Real.*

What is now known as Walt Disney Imagineering began in 1952 as a small team hand-picked by Walt Disney to make his dream of a magical kingdom come true.

Today, this creative think-tank and development subsidiary of The Walt Disney Company includes a team numbering more than 1000 who support the design, creative development, execution, and ongoing show quality of 11 theme parks on three continents, 36 Disney hotels and resorts, Disney cruise lines, and the dining and entertainment venues of ESPN Zone®. Although Imagineers don't run the parks on a day-to-day basis, they maintain a continuous involvement as the parks grow and evolve each year. Imagineers translated Walt Disney's original vision for Disneyland and remain equally active in keeping the park fresh and original more than 50 years later.

True to the culture. Imagineering's corporate culture, while built on storytelling, making magic, and doing the impossible, also incorporates a delicately balanced strategic discipline. This combination yields an organization capable of sustaining and expanding a set of physical and brand assets that have taken decades to build and that must be nurtured and maintained for many decades to come.

Despite the growth in scale and scope over the past half century, Imagineering remains close to its cultural roots. Whether making sure that park guests are greeted at the gate by a fully animated, interactive Disney character or sending them on a high-speed thrill adventure through the Himalayas with an inevitable encounter with the legendary Yeti, Imagineers never cease to take great pride in "doing the impossible."

Combining the talents of more than 140 disciplines—model makers, software developers, artists, sculptors, writers, engineers, architects, music experts, special effects designers, project managers, film makers, scientists, animators, and landscapers to name only a few—Walt Disney Imagineering offers the consummate example of successfully merging creativity, technology, and business savvy. Through this process, Imagineers bring to life the wonderful magic of Disney around the world.

Once Upon a Time

* ✳ *

All Imagineers—the term they proudly use to self-refer—can tell the story of Disneyland's genesis. As the story goes, on Saturdays Walt Disney would sit on a park bench while his daughters rode a carousel nearby. Walt Disney felt there should be some kind of an amusement enterprise where parents and the children could have fun together. And that's how Disneyland started.

Tom Fitzgerald, Executive Vice President and Senior Creative Executive, adds some insight to the story. Disneyland was a natural progression for Walt, who had already had great success transforming static pictures from storybooks into talking, dancing, singing characters through animation in his movies. It must have seemed just a logical progression to transport the active two-dimensional figures of the movies into a three-dimensional space where people—his guests—became active participants in the stories and the experience themselves.

Not everyone fully appreciated Disney's concept, at least not at first. According to the "story," when Walt Disney told his wife about his vision for an amusement park based on the magic of storytelling, she dismissed the idea. Amusement parks were tawdry and dirty, certainly not the kind of place for wholesome, family entertainment. Walt countered, maintaining that things didn't have to be that way.

Walt Disney had a very different view. He wanted to do the impossible, to invite his guests to step into the story and to be immersed in every sense by the experience. That was the vision that was realized in Disneyland, and that view of "doing the impossible" not only set the tone for the original hand-picked team of film writers, directors, and designers who made Walt's dream come true, but also for every Imagineer since.

Veteran Imagineer Martin A. Sklar, Vice Chairman and Principal Creative Executive, has said that "Imagineers turn the spark into a park" taking an idea and then applying their own specialized expertise to literally bring a story to life. "Anyone can dream," he said, "but Imagineers can make a dream come true."[2]

The Story of Innovation

* ✳ *

Storytelling, the heart of the Disney enterprise, is also the first step in any innovation project at Imagineering. Just as Walt Disney

---

[2] Martin A. Sklar, Vice Chairman and Principal Creative Executive, Walt Disney Imagineering video presentation.

used pictures to sell the Disneyland vision to his bankers more than half a century ago, Imagineers start every project with a "storyboard" or some form of artistic rendering of the concept (see Figure 8.1). In fact, according to Imagineering corporate lore, Walt Disney invented the storyboard, a technique that virtually every motion picture director uses in developing a movie concept.

Tom Fitzgerald says, "We're totally influenced by the same path Walt Disney took. We storyboard almost everything. It's a way to visualize a movie, a park, or an attraction before you make it. To show the vision before you get started. At that stage in the process, it doesn't cost any more to sketch something incredible—like a monkey that talks—than it does to draw a chair."

According to Fitzgerald, the storyboard is a good way to get everyone on the team "to understand the story we're trying to tell." It helps people look at their jobs as part of the storytelling. For instance, the landscaper who creates a frontier town in the 1800s uses flowers and trees as another dimensional tool for storytelling."

The magic of innovation. Although few companies can claim a comparable record of sustained innovation over a span of fifty years, Imagineering does so without a formal, structured innovation process. If asked to describe the product development process, an Imagineer does not pull out a procedure manual. Instead an Imagineer refers to—what else?—a storyboard hanging in a hallway at company headquarters (see Figure 8.2).

Like any good story, the innovation process ebbs and flows with characters—the development team members—who move in and out of the script. Although, predictably, all projects are expected to end "happily ever after," no two stories follow the exact same path from their original storyboard to their final

Figure 8.1 Walt Disney's Original "Rendering" for Disneyland, used to "sell" the concept to his bankers. © Disney.

Figure 8.2 The Imagineering Product Creation Process. This wall-size storyboard depicting the Imagineering product creation process hangs in a hallway at company headquarters. © Disney.

credits. All, however, face the same key points of suspense in the concept approval and capital appropriation process, but at their own pace—as befits any well-told story—not according to some predefined timeline.

For example, the story of the new Disneyland theme park in Hong Kong began as speculative conversations in the mid 1990s. It wasn't until 1999 that the funding for the theme park and the budget for the resort's hotels and infrastructure were approved by the Disney board of directors. More than six years in development, the park opened for guests in 2005. Like the original Disneyland, which celebrated its 50th anniversary in 2005, the expected "product life cycle" for the Hong Kong park is measured in decades, not weeks or years like consumer or electronic products. Planned, continuous renewal and reinvention ensure that such major investments reinforce the Disney brand and achieve the desired financial returns. The Hong Kong SAR Government estimates that the first phase of the project will generate a present economic value of HK$148 billion (US$19 billion).

Strategic Discipline

* ✳ *

Innovation on such a grand scale without a highly formalized new product development process provides further evidence of the magic that Disney and its Imagineers create. But Martin Shumaker, Vice President of Finance, likes to put a bit more structure around innovation. He explains that there are basically two types of innovation processes.

The first he calls "true blue sky." This, he says, can be based on practically anything from "an exciting new technology that we've seen somewhere" to a "fun cool idea some writers or concept designers have been thinking about" for a new special effect, a new ride, or even an entire new theme park. "There are few limits," he says, "when we're talking true blue sky" (see Figure 8.3).

The second type of innovation process, called "targeted blue sky" evolves from an identified customer need, for example, when one of the theme parks seeks a new attraction. The need could be based on guest research that says that the park may not offer enough entertainment for very young children. Or it could be based on observation or guest statistics that indicate underutilization of a certain part of the park. The targeted blue sky process

Figure 8.3 An Imagineering Blue Sky Session

would then work to fill the identified need, as creatively and magically as possible of course.

Don Goodman, President of Imagineering, elaborates. "Probably half of our great ideas have been things that no one had ever asked for, the true blue sky ideas." He says that he sets aside a certain budget—the Blue Sky Fund—that he divvies up about one-third for true blue sky and about two-thirds for targeted blue sky concept development.

Shumaker is quick to point out, however, that even the most highly innovative concepts incorporate more strategic discipline today than in years past. He says, "Walt Disney Imagineering was originally set up as a think-tank: build it and they will come. If you could pitch the creative concept and get executive buy in, someone else would figure out the rest later.

"That was 20 years ago. Today, we're working more on marrying the creative process to business planning—high concept approval now has some very well thought out business decisions associated with it. We're getting the finance people more involved. To sell the projects we have to discuss business, but we do it in a positive way. We don't want to infringe on the creative process, but we also have to bring a business sense to it—a long-term perspective guided by the brand, just not in a heavy handed way."

Shumaker points to the new Expedition Everest attraction at Disney's Animal Kingdom in Florida—where guests go trekking through the Himalayas only to encounter the savage Yeti—as an example of the integration of creativity and business planning (see Figure 8.4). That project, he said, came out of "true blue sky thinking where we were just looking for new ideas. But once we've got the idea, it then can take six months to a year just getting the 'green light' on the concept. And only after the parks have

Figure 8.4 Expedition Everest. True Blue Sky Comes to Life, exemplifying how Imagineering integrates creativity and business planning. © Disney.

said they want it, does it become a real plan, tightly budgeted, fully staffed, with a well-defined timeline."

Don Goodman adds, "We have a mixed bag of linkages between new product development and business planning. We can generate intellectual property—our creative concepts—without having direct linkage to business planning on the front end. That's the kind of magic that just sort of happens. But once you create that piece of intellectual property, putting it through the systems and processes to bring it to life is very rigorous, and in some ways much more traditional."

A full measure of magic. Besides budget, staffing, and scheduling, Imagineering looks for other ways to measure the success of their innovation processes. Shumaker says that because "we're always working on how to make magic in lots of different ways," we look at the quality of the guest's experience very carefully. Says Shumaker, "Our guests are very passionate, so they speak up, often without even being asked. But we also do formal guest satisfaction measurement regularly on all new and existing attractions."

Entertaining the Idea of R&D

* ✳ *

The words research and development usually conjure up visions of high-tech laboratories, engineers of all stripes, and a singular focus on tangible, physical products. Perhaps that's why Bruce Vaughn, Vice President of Research and Development, says his group's very existence at Disney is an industry anomaly. "No other entertainment group has an R&D group like us," he says, but quickly adds, "And we don't look like any other R&D group anywhere else. We do directed R&D toward an end that meets a unique need of the company."

The Magic of Intellectual Capital[3]

According to company released statistics, Walt Disney Imagineering "has been responsible for approximately 115 Disney-owned patents" in areas as diverse as:

---

[3] *Walt Disney Imagineering Fact Sheet*, June 2003.

- Ride systems

- Special effects

- Interactive technology

- Live entertainment

- Fiber optics

- Advanced audio systems

He also notes that the mix of people in R&D is unique. "We're sort of a separate subset of Imagineering, a very small group of only about 40 people." The key to sustaining a high level of R&D, not only in creativity but also in sheer output, he says, is casting—the term Disney uses to refer to how it recruits and hires staff.

"You have to get the right people and the right mix. We need to take an idea from a concept to its first representation in a very short time frame. The mindset of the people who do that has to be to mock up something quickly, learn what doesn't work, change it, test it, then go back and do something that does work."

To accomplish that, the R&D group includes an unusually diverse mix of artists, scientists, designers, writers, various engineers, and fabricators. Vaughn maintains that each member of the R&D team "has to be world class, because there are so few of us." He also says that every team member must "be a sort of Swiss army knife" so each can flow in and out of various areas of expertise depending on the project.

"They must have that passion for inventing the new and a tolerance for compromise, for finding the right balance between the creative concept and its technical execution. Some people don't like that. Our people must." That's why he says he "casts" all projects with both creative and technical members, to enable an iterative process. "It might be a creative idea that starts the ball rolling, then the technical person comes up with a way to enable that, then the creative asks for a bit more. It's a zigzag, a good back and forth."

Bruce Vaughn can speak about this process with great authority. It's part of his own personal journey to the top of Disney's R&D group. An undergraduate English major with aspirations to go to film school, he bowed to family pressure to go to an eastern law school instead. Dissatisfied, he left after just a year, and through a series of unusual "career moves" (including pool cleaning and truck driving) soon found himself working for a movie special effects studio. That led to more experience in the technical side of movie production, an eventual move to California, and a "mind-blowing" trip to Disneyland that first, and forever, ignited his passion for making magic, the same passion with which he now inspires his R&D team.

Vaughn traces his success at Imagineering to his diverse education and experience. "I came to the table not only with an education in literature and storytelling, but also with a technical background and work experience. I spoke both languages, so I could help make sure the creatives and the engineers understood each other from the start and didn't go off in totally different directions. I became a translator, a liaison between the technical and the creative side. Then, over time, through a series of jobs and bosses, they made me the head of the group." It should also be noted that Vaughn's success stems not only from his ability to speak both languages, but also his strategic vision, which as he puts it, is "to push the edge of the envelope in advancing the art of storytelling."

The Art of Casting

* ❋ *

Bruce Vaughn's career at Disney is just one of many examples of the importance of "casting"—hiring the right person for the right role at the right time. And one of the most common and distinguishing characteristics among Imagineers is the passion they bring to their work.

Take Tom Fitzgerald, Senior Creative Executive, for example. He traces his acute fascination with all things Disney back to his childhood introduction to the "wonderful world" through television, and even more specifically to the 1964 New York World's Fair where his visit to the Disney exhibit cemented his decision to one day work at creating just that kind of magic.

Says Fitzgerald, "When I was in college at Northwestern in theater communications, I figured that if I was really serious about a career with Disney I needed to know how things work there. I landed a job as a ride operator at Walt Disney World in Orlando so I could observe and take careful note of what guests liked and didn't like. The second year, I conceived, designed, and built a model for a new theme park attraction out of foam core and paper. I put it in the back of my car and drove to California." While the attraction Fitzgerald designed never made it to a theme park, he says that Disney's willingness "to take a chance on a guy for his dream job" is totally consistent with the company's commitment to "cast" the right combination of passion and talent for the task.

Bruce Vaughn agrees on the necessity of passion for the job, but as the head of the relatively small R&D group he looks for other qualities as well in his casting process, especially in selecting the members of what he calls the next generation of Imagineers.

"When we're staffing for a new project and find someone who shows great promise, before hiring we like to take a test drive to see if that person can fit into this culture, this tribe. We tend to bring them in as interns or consultants first. If that test run goes well and the project gets the green light, we'll bring them in as full time people."

Vaughn says, however, that when he's replacing a core team member, casting becomes even more critical. He says he often finds the right people within the company. "R&D tends to have people who didn't work out particularly well in other groups. People who had big ideas and were outspoken. They might not fit in another group's culture, but we encourage that. We're looking for that different type of person."

Fanning the flame. Don Goodman acknowledges that R&D requires a very special personality indeed, but also notes that the challenges and nuances of managing creative people extend throughout all of Imagineering. Says Goodman, "I think there's an art to it and despite the fact that I used to be a CPA, I've gotten better at it over the years."

He compares his role to that of a business manager for the symphony. Everyone wants the symphony to make money, he says, but the musicians are really just interested in performing. "Working here can sometimes be an emotional roller coaster. So much personal identity gets wrapped up in what they do that you must be careful not to crush their spirit."

That's why he says, "You'll never hear someone get slammed at a meeting. Of course, some projects are just not right for the time or place, but we never throw anything away. You won't hear us say that something will never work. After all, our job is to do the impossible."

Goodman also notes that in a company where an individual may not actually see the results of their work materialize for several years, personal recognition becomes very important. "For a group that gets mostly team recognition, there's a lot of value in personal recognition—a note from Bob Iger or simply a guest comment being passed along. A huge chunk of what I do is keep people from feeling underappreciated."

Passport to Tomorrowland

* ✸ *

So where will Imagineering make its magic next? Sue Lucas, Vice President International Development for Disney Parks and Resorts, says that although Disney has been building parks in locations like Paris, Tokyo, and Hong Kong for decades, global expansion was not a separately defined responsibility until very recently. Her group, she says, is now solely focused on getting international deals done.

Says Lucas, "It's not just finding markets where the Magic Kingdom works. We also have a wealth of other products we can market internationally. Because of cultural differences, brand affinity, or even geopolitical stability, our answer for a specific geography might very well not be a theme park. So we look more broadly for that product or idea that will get us into a particular market."

While consistent global brand awareness remains a priority, Lucas says the company doesn't always have to tell exactly the same "stories" everywhere, every time. She says, "What we always ask ourselves is: what is the problem we're trying to solve? It's the same as when Walt Disney didn't just want to watch his daughters play, he wanted to experience it with them. It's difficult to innovate if you're not trying to solve a specific problem, and in different markets the problems are different."

Don Goodman echoes much the same sentiments about international expansion. "Some parts of the world don't know the Disney stories. Some people don't know Peter Pan, or even Walt Disney. If there's no distribution of the stories through the Disney channel or consumer products, you can't lead with a theme park. That's why we can't be separated from the rest of the Disney company, the distribution channel. It's why we've partnered very closely with the other business units."

Can Magic and Discipline Co-Exist?

* ✳ *

The product creation processes at Walt Disney Imagineering may appear quite removed from traditional notions of what happens in new product development elsewhere. But why wouldn't making magic require a distinctive approach? That said, Imagineering adheres much more closely than one might think to many of the product creation principles described in this book. They just do it in their own special way.

For example, at first glance one might question the degree of discipline and rigor in a product creation process that is documented only on a storyboard. Clearly, creativity is the mantra, but a second look reveals that everyone at Imagineering knows the company's product creation process chapter and verse. It's indelibly ingrained in their culture and their common understanding of it is the foundation for the strategic discipline that marries creativity to business goals and turns true blue sky thinking into revenue generating projects.

Imagineering takes the voice of the customer very seriously, but more as a wide-angle lens through which to spark creativity than as a lens to focus narrowly on a pre-defined need. Often insight comes just as much from what park guests don't say as from what they do. And the company's willingness to allocate significant resources to true blue sky projects underscores that while it is important, customer research and feedback don't impose the kind of constraints on product creation that might be found at other companies. They actually liberate the process.

To achieve Imagineering's level of process understanding, creativity, and strategic discipline requires a very special and diverse mix of people. That's why, true to its culture, the company refers to recruiting and hiring staff as casting. It makes sense. With so many different skills, talents, and abilities required to tell every story, putting the right person in the right role at the right time is absolutely critical.

The Next Great Thing

* ✳ *

Looking broadly at what lies ahead for Walt Disney Imagineering, Don Goodman says, "Most corporations want everything to be controllable, for the future to fit neatly into a box. That's not possible here.

"Many if not most of the things we try have never been done before. I tell our R&D people that they are required to take risks, but the risk has to have the potential to become something great. And at the corporate level, although they know we have a diverse menu of projects always in development, I always remind them that because of who we are, we need to do something every few years that is absolutely groundbreaking. I'm very pleased that our management really understands that: the relationship between how much risk you take and how great the payoff is."

Chapter 9

MARS INCORPORATED

———— ✳ ✳ ✳ ————

Consumer-Driven Innovation

The intoxicatingly sweet smell of chocolate wafts through the air in the parking lot of Mars Incorporated, set in the rural town of Hackettstown, New Jersey, 50 miles west of Newark. Despite the North American headquarters' unpretentious exterior, glass display cases in the lobby remind visitors that Mars owns category-leading brands instantly recognizable throughout the world: snacks including M&M'S, Snickers, Dove, Starburst, and Skittles; pet food brands such as Pedigree, Whiskas, and Cesar; and "main meal" products under the Uncle Ben's brand.

Through security and up a wide flight of stairs, one finds a large, open room where hundreds of employees sit at modern workstations organized in pods unobstructed by cubicle walls. In their center sits the president and the executive staff of Mars Incorporated's North American snack business. This privately-held, multibillion dollar enterprise breaks many molds.

Private, but Global

* ❋ *

Mars Incorporated traces its roots to the early twentieth century and a pair of father and son entrepreneurs. Frank Mars, the patriarch, entered the candy business, around the turn of the century, as a wholesaler of penny candy in Washington State. By the early 1920s he had established himself as a small-scale manufacturer of candy.

According to company legend, Frank and Forrest came upon the idea for the Milky Way® bar while drinking malted milks at a local Chicago drugstore, and the candy bar, with its malt-flavored nougat center and layer of caramel surrounded by milk chocolate, became an instant sensation. Inspired by his father's success, Forrest enrolled at Yale to study industrial engineering and the world of commerce.

The two continued their collaboration by building a state-of-the-art manufacturing plant in the Chicago suburbs, which churned out 20 million candy bars a year by 1929. Further growth came with the introduction of the Snickers bar in 1930 and the 3 Musketeer® bar in 1932.

In 1933 Forrest headed to England with plans to build a business empire. He adapted the Milky Way formula to British tastes and renamed it the Mars® bar. It quickly made headway against the then dominant British candy companies of Rowntree & Co. and Cadbury Brothers Ltd. Intent on a broader empire, in 1934 Forrest bought a small British company that canned meat by-products for dogs, a novel idea at a time when most pets simply ate table scraps. Forrest, the innovative salesman, saw the potential to sell pet owners on the nutritional benefits of canned food. Sales of Petfoods Ltd.® grew fivefold in five short years and became the foundation of today's global pet care business at Mars Inc.

As World War II broke out in Europe, Forrest Mars returned to the United States to further expand his empire. In 1939, Forrest Mars formed a business partnership with Bruce Murrie, the son of the Richard Murrie, longstanding president of Hershey, the other leading candy maker in the United States. The candy-coated chocolates from M&M Ltd proved another innovative success from the Mars family. In 1942, Forrest expanded his product range and empire yet again by buying a rice mill that offered a non-sticky, more nutritious product through "parboiling" or steaming the rice while still in the hull. Forrest came up with the idea for branding "Uncle Ben's® Rice" over lunch in Chicago with his advertising firm. Today, the brand undergirds the Main Meal Food division of Mars Inc.

In December of 1964 Forrest Mars combined his empire with that of his father's and renamed the $100 million holding company Mars Inc. Forrest continued to build his global empire until the fall of 1973 when he turned over control of the company to his three children, Forrest Jr, John, and Jacqueline, all already engaged in various ways with the business. This third generation of Mars family members took the company from $300 million dollars in revenues to the current level of over $18 billion worldwide.

In addition to bringing their own unique contribution to the management of the company, which now spans more than 100 countries, the current generation also codified the "Five Principles of Mars," which capture their father's management philosophy:

- Quality: The Consumer is our boss, quality is our work and value for money is our goal.

- Responsibility: As individuals, we demand total responsibility from ourselves; as associates, we support the responsibilities of others.

- Mutuality: A mutual benefit is a shared benefit; a shared benefit will endure.

- Efficiency: We use resources to the full, waste nothing and do only what we can do best.

- Freedom: We need freedom to shape our future; we need profit to remain free.

Consumer-Driven Innovation

* * *

Despite a long record of successful innovation and growth, even Mars Inc. has occasionally stumbled in new product development. By the mid-nineties, Mars resolved to sharpen its innovation edge. "The 80s and 90s were mostly about geographic expansion," explained John Helferich Vice President of University Research. For example, Mars introduced their products to Muscovites in 1990, around the time of the collapse of the Soviet Union.

By the summer of 1995, the company had opened a 100 acre plant capable of producing 70,000 tons of candy per year. By the end of the decade, Mars had become the dominant Russian candy maker with a market share of 40 percent.[1] Similarly aggressive expansion placed Mars as the number one candy maker in the world at the turn of this century.

Until very recently, according to Doug Milne, Senior Marketing Manager for the M&M'S candy brand, the company had focused mostly on a product platform approach to innovation, testing new varieties of M&M'S. "But that wasn't a consumer driven strategy," says Milne in describing a watershed change in Mars' approach to new product development, which shifted them from iterative product extensions to consumer focused innovation based on marketplace research. Milne cites as a prime example an ethnographic research project that focused on how M&M'S fit into the snacking behaviors of consumers. Through that research, the company began to view snacking as occasion based. That, in turn, led to a number of insights and innovations.

Gifting and special celebrations was one of those occasions where consumer insight led to marketing innovation. Research showed that consumers responded well to having a choice of colors for their M&M'S. For example, they wanted to match the colors of a bridal shower or to only have the candies in the colors of their favorite sports team. In response, Mars now makes M&M'S available in 21 different colors, which can be ordered over the Internet and mixed and matched in any combination and quantity to suit the consumer's preference. That proposition proved

[1] *The Emperors of Chocolate: Inside the Secret World of Hershey and Mars*, Joel Glenn Brenner, 1999, Random House, New York.

so popular with consumers, that the company now offers M&M'S with personalized printing of initials, names, and messages on the individual candies.

Mars hit the sweet spot when they combined well-honed consumer insight with their traditional approach to platform expansion to develop and introduce the Snickers Marathon Energy Bar®. Having been interested in the energy bar category for some time, in the early 90s, the company entered the limited but growing market with a bar called VO2 MAX. The product was targeted at athletes to enhance rapid muscle recovery following exercise. Mars withdrew the product, however, because sales did not meet expectations.

Drawing on the valuable technical and marketing lessons learned from that experience, Mars went back to the drawing board. In 2003, the company introduced a revamped product the Marathon Energy Bar, under the Snickers® brand umbrella. Within one year, the two versions of the bar had become the two fastest turning products in the category.

"This is such a great success story," explains Bob Boushell, Innovation Portfolio Director. "We had watched the energy bar market grow over the years and really wanted to get back in the game. We knew that we could leverage our technical and manufacturing platform expertise in bars. The tipping point came when larger retailers began to establish energy bar sections in their stores, which put the business within reach of Mars sales force," Bob explains. "The challenge then became how best to get in and win."

At almost $1 billion in retail sales, the energy bar category had only penetrated 15 percent of U.S. households at the time. Mars knew that its entry should be credible and accepted among current energy bar consumers, but most important, become the bar of choice for new consumers, the 85 percent of households not yet purchasing in the category. Digging deeper into sales data, in virtually every sub-segment, chocolate and peanut butter were the top selling flavors. With its equity in the Snickers brand using chocolate and peanuts to satisfy hunger, the company determined that the brand would be the vehicle to re-enter the energy bar category and to make it a mainstream product.

Consumer research and consumer centered design were the key factors in bringing the product successfully to market. Mars knew that the primary barrier to usage for the 85 percent of consumers they had targeted was the unacceptable taste of existing products. They thought the idea of a portable, nutritious snack fit their busy life style, but were unwilling to sacrifice taste for the convenience and health considerations. They also found the appearance of most bars on the market unappealing because they couldn't visually recognize any familiar ingredients in the bars. Digging further, Mars found the proper macronutrient balance that was essential to delivering a credible product.

Combining research and consumer insight, Mars had found its formula for success: Snickers brand name, great peanut and chocolate taste, recognizable food ingredients, and the proper macronutrient balance. One challenge remained: the company's food scientists had to break through and deliver.

Mars' "sensory team" identified the major components of the competitors' off-taste and dry texture. Doing what Mars has mastered—making great tasting snacks—the company used prototypes in repeated consumer testing to refine the product concept. Both the product formulation and the manufacturing process contributed heavily to superior taste and texture delivery.

In final testing, the products got a "big wow" among the targeted consumer audience.

Scaling up and launching quickly proved the next major hurdle. Knowing the market was already primed and the opportunity space was limited in terms of timing, Mars modified an existing production line and began turning out the bars as quickly as possible. Now with market success confirmed, the company has gone back into innovation mode to add greater efficiency to the process. Says Boushell, "We know this will continue to be a very active and competitive category. Our plan is to keep the innovations coming, both on the store shelf and in the factory."

In a more tightly focused consumer research study, the company wanted to find new ways to meet more of the consumer's needs at the Easter season. Although Mars uses focus groups frequently, they felt that in this instance better insight would come from actually observing how consumers use candy and snacks during the Easter holiday.

The researchers gave cameras to consumers and asked them to photograph the process of how they put together their Easter baskets initially and then to continue to record how the baskets appeared during the following days. Next they invited the participants to come in and describe their experiences as documented in the photos. The quality of information gathered was much richer and more detailed than would have been gathered in a traditional focus group. It provided the insight that will influence future Mars' Easter candy offerings.

Expanding creativity and consumer understanding. Although the Sensory and Consumer Research Group often works closely with market research, they draw a clear distinction saying that the group's primary purpose is to support R&D. The group's three main areas include sensory evaluation, consumer insight, and prototype and design.

Mars R&D feels that sensory evaluation and consumer insight must be closely aligned. Consumers can't articulate the qualities of the product in a way that is useful to a product developer. Wanting "more chocolate" often means they want different texture or more milk flavors. So Mars employs a rigorous method to describe a product in terms of what's there and what the intensity is. Think of it like a wine taster, only for chocolate. Mars has people who can describe chocolate flavor and intensity in at least 15 terms. But it's also very important to have sensory and consumer people working together on the whole product.

One example is a product that consumer research said needed to be lighter to eat. Changes were made, but the product still failed to perform up to expectations. Further investigation found that Mars hadn't changed the attributes that were meaningful to the consumer. What the consumer was really saying is that there's a lot of work involved in eating the product. Tasting experts ultimately translated that in a term that R&D could understand much more clearly: cohesiveness of mass.

The Sensory and Consumer Research Group also includes the prototype and design team. The thinking is that a prototype is a communications vehicle. Mars researchers note that when trying to describe a new idea to someone, whether it's a consumer or our own internal decision makers, it's much easier with a rough prototype. It doesn't have to be 90 percent perfect, just good enough to get the idea across. The prototype offers a springboard to develop the idea further based upon the feedback.

Prototyping helps to stimulate creativity and original thinking. Mars insists that its people should not feel constrained by the limits of what can be done today. The Sensory and Consumer Research Group challenges the organization. How far can they push the Snickers bar? What else could Mars do with peanuts, caramel, and chocolate? Or maybe choose an innovative ingredient and try to put it in as many different formats as possible.

Prototyping and design draws on diverse disciplines, both inside and outside the company. Mars has chefs and food scientists in house, but also brings in the artistic view from the broader culinary world. They bring in world famous chocolatiers as consultants and tap into the staff from the Culinary Institute of America to help keep current on the latest food ingredients and food trends.

Transforming Pet Food

* ✸ *

A powerful example of Mars' transformation to consumer-driven innovation can be found in a new product recently introduced by the pet care division. Mars set out to transform main meal feeding of dogs and achieve competitive differentiation from everything currently available. The objective grew from ongoing ethnographic research of what Jerry Franks, Innovation Manager at the Petcare division, describes as "compensating behaviors and unarticulated needs."

Franks explains, "There wasn't much enthusiasm with feeding. It's messy for the owners, and the dogs wolf down the food in about 30 seconds. We used a dedicated multifunctional team headed by a marketing guy and a full-time research & development person. R&D focused on the product and packaging while marketing concentrated on the business case and the sales story."

The team included other full-time people from supply and finance plus part time assignees and outside resources as needed. Over the course of the project, the team conducted market research including pet feeding panels with the help of statistical analysts, vendors, dedicated pet specialists, innovation management experts, and project management practitioners.

From idea to product. To help the ideation process, Mars partnered with a well known design house that has helped stimulate innovation for a wide range of products and companies for more than 25 years. "We considered a host of ideas like dispensers for kibble or using pouches instead of cans," explained Franks. "But the idea that really stuck was the concept of a 'bucket of bones', like at Kentucky Fried Chicken, but full of artificial bones. That was the start."

"This was a great example of purposefully breaking category conventions and norms," adds Chris Jones, Marketing Vice President for pet care. "Usually we look for bigger, higher, faster, but we decided to take a completely different tack. Eating out of a bowl is not a normal way for a dog to eat. Dogs like bones."

Inspired by the idea, one of the Mars team members made a prototype at home over the weekend. "Ninety five percent of us are dog and cat owners, so we can look at how products perform very quickly," explains Franks. "It's a really good, quick-and-dirty way to test an idea. We found that the dogs we tested took the bone out of the bowl by mouth and held it with their paws to eat, which the owners loved. The product we started with didn't work very well. The attributes were wrong: texture, taste profile, size. But the idea was right."

Encouraged, the team set out to conduct rigorous analysis to determine the right characteristics from both the dog's and the owner's perspective. As a main meal, the product, now known as WholeMeals™, needed to meet certain nutritional requirements. In pet food, the nutritional characteristics encompass the full "lifecycle" of the food. "We also have to design for what comes out the back end," Franks explains sheepishly. "The consistency and quality of the stool must be acceptable as well as the smell."

The team also looked at perceived enjoyment from both the owner's and the dog's perspectives. Owners were asked to rate various prototype designs, and dog preferences were measured by how much they ate and their choices in side-by-side offerings. "We spent six weeks with dog owners to refine the attributes," Franks said. "We watched them and talked to them. It was a very intense six weeks of iterative testing. One concern we never hear is that the WholeMeals™ product is messy. When people use it, they discover it's less messy than wet food.

"Early on we took it to in-home testing. Researchers videoed the product in action each day to discover how engaged the dogs and their owners became with the product. After the trials, people asked for more product samples. That was another great indicator that it was going to be good.

"We delivered the 'cost of entry' attributes, and are beating the conventional meal products with greater dog enjoyment, nutrition, plus dental and oral care. It's the first truly complete meal. Complete in nutrition, enjoyment, oral care. With kibble you have 60-second consumption experience with little pleasure by the dog or the owner. Eating WholeMeals™ takes 6 to 8 minutes which provides enjoyment for owners and their pets."

Process innovation. But an innovative product design does not guarantee success. The Petcare division needed to produce the product at a competitive cost. One way to keep cost down was to use existing technology and avoid new capital investment.

WholeMeals™ is an extruded product similar to other pet foods, but with two flows that come together to create a hard core and a soft outer shell. It looks something like a candy bar (see Figure 9.1). Modification of existing equipment by process engineering and supply personnel increased production rates fourfold during the development phase, which dramatically improved cost.

Packaging also presented an opportunity to leverage corporate-wide expertise while adding further advantages for the retail customer. "Because we're turning main meal feeding on its head, we're doing different packaging," said Franks. "The Whole-Meals™ is individually wrapped in foil which allows us to pack it in a box with a handle for ease of carrying. That packaging design appeals to the retailer because boxes pack out better on the shelves, are easier to handle than bags, and convenient in high speed warehouses."

Although a transformational approach to main meal pet feeding, such innovative packaging was easy to handle at Mars. "A carton is a new asset in main meal dog food, but we're a multi-product company," notes Franks. "There is huge synergy across the corporation. We know about collating and packing individual items in a carton. We have those technologies already

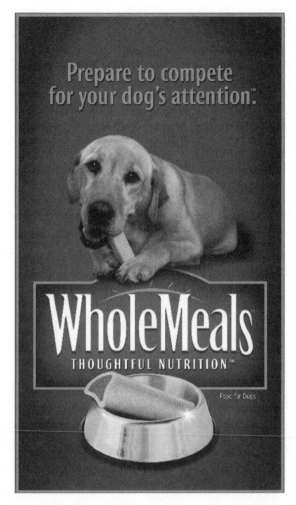

Figure 9.1 WholeMeals™ by Pedigree. WholeMeals' bar shape proved best in the testing cycle and the individually wrapped bars can be packed in a box with a handle for ease of carrying

in our company." The innovative product leveraged existing corporate assets to create great "ease of use" advantages for consumers and retailers.

Selling radical innovation. But the innovation doesn't stop with the product design and the supply process. "One of the project themes would have to be that it's not just the product; it's the business model, the consumer mind set, and the marketing piece," explains Franks. "Innovation is at its core, but innovation in various elements is what makes this successful."

"We've deliberately blurred the distinction between main meal and treats," said Franks. "So, positioning—main meal or treat—was the decision we needed to make. We have a name that conveys meal, we've added feeding guides to show the equivalency of the WholeMeals™ with canned dog food and dry dog food. Although it can be initially confusing to the owners, after several exposures to the product 80 percent of the people in our tests say they would use this as a meal."

The Mars team has analyzed the effects at first exposure as well as ongoing feeding. "We have to take both attributes into consideration," explained Franks. "We wouldn't want something that took two weeks to get acceptance. We need quick acceptance and long term usage."

"Pricing has also required innovative thinking," said Jones. "Consumers have different expectations on pricing. We've got a somewhat premium pricing to match the superior product attributes. The price is appropriate for the unique product benefits and supports the market positioning."

The company is also employing an innovative approach to distributing WholeMeals™. Distribution started with a small group of specialty stores which have the knowledgeable sales clerks who can explain the new product concept. This initial "viral marketing" approach will help Mars fine tune the product and message prior to mass-market distribution in retailers such as

Target, Wal-Mart, and Kroger®. That channel expansion will also require a change in the marketing approach to mass advertising.

"It's a step function," says Franks. "It is a disruptive innovation which will have a tipping point. What does it take to become critical mass or mainstream? That will only be understood when it happens. I could have this business that creeps long with small penetration, then suddenly hits the tipping point and explodes. As a privately-held company, we're not tied to this quarter's revenue or earnings from a Wall Street perspective. We're willing to spend the capital in the upfront launch and patiently wait for the success point."

Process Discipline and Connectivity

* ✳ *

Mars Research & Development works to routinize such innovation to ensure a consistent portfolio of new products. "To maintain a better control of the process, we have instituted a formal stage-gate and portfolio management process," explained Helferich. The portfolio process starts with a segment review. "It's a high-level meeting to show a five year view of the business, innovations that will take place, elements that will drive the business, the breakthrough things" explains Doug Milne, Senior Marketing Manager for M&M'S. Mars then translates the five year plan into annual plans on a rolling basis finalizing the details as each year approaches. Milne continues, "You'll pick a target year for each innovation in the five year plan, get general thoughts from customers, but won't lock anything in until 18 months before the year in question."

To translate the five-year plans into an appropriate portfolio of projects for implementation Mars applies a systematic, three-step process:

1. Strategic Screen: Does it make business sense?

2. Development Cycle: Is the idea ready for market?

3. Market Cycle: What do we need to make it, sell it, take it to market?

Brian Gracyk, Business Planning Manager for Snackfoods, elaborates "Every four months we have a strategic review for development proposals. We try to roll all of the phases into the same period review."

To ensure a balanced portfolio, Mars plots the projects on a two-dimensional chart. The horizontal axis measures the technology reach and the vertical axis shows the market reach. The size of the bubble used for each project indicates the level of resources currently applied. A balanced portfolio has a clear mix with a thoughtful allocation of resources based upon the real business potential.

"Senior management must approve the specific projects to get resources," explains Steve Crawford, Snackfood Industrial Engineer. "If a project doesn't get senior level approval of an idea, the project doesn't happen. Once it's a big project, not just an activity, it gets a project manager who reports progress to date on a regular schedule."

The innovation portfolio is not limited to product innovation alone. Bob Boushell, Innovation Portfolio Director, highlights the role of process innovation at Mars, "We do as much process innovation on products in the marketplace as we do product innovation for new products we want to introduce. We have a long history of process innovation behind the scenes on our existing brands like M&M'S and Snickers. The consumer sees the same product and the same quality, but behind the scenes, the factories can be substantially transformed."

Boushell notes that process innovation takes longer, "If we have a great new product concept, we'll get the product out to the consumer quickly using existing processes and then do the process innovation behind the scenes. We probably do more step function changes in production processes without changing the product. For a new product you need to get to market, you might wait until later for the next step function change in process engineering."

David Pontzer, Group Manager Innovation Engineering, describes the challenge of balancing product and process innovation: "A large part of my job is to decide when to plug the process technology development into the critical path. Another consideration is the need to leverage conventional assets in case the sales of the new product don't ramp up as fast as you expect." Myriad details must be considered in managing each project to capture the greatest business opportunity.

Doug Milne makes the important point that not everything in the innovation portfolio should be a break through, "We also tweak the core brands: new product development and new programs alone will not drive the business over the long term. You have to refresh the existing products every couple of years.

For example, we stepped back a couple of years ago and asked if the M&M'S brand was living up to its brand essence and concluded that it was a bit tired. In January 2003, we transformed the entire product line of M&M'S to black and white for a 10 to 12 week period, and then re-launched the whole brand with new colors and fresh packaging." Such changes keep the product and message current even though the basic product and the essence of the brand remain unchanged.

The black and white M&M'S project may not have affected the essence of the brand, but it did require an enormous level of cross-functional integration. Steve Crawford, Business Development Team Industrial Engineer, explained the complexity, "It was a huge logistics challenge to manage the whole pipeline of product from suppliers to manufacturing to distribution to the retail customers. We had to watch inventories for old colors and for the black & white packages to limit overlap with the new color."

Bob Boushell reinforces the need for effective execution through cross-functional collaboration, "When you think about how much we invest and how little we charge you can see the need for speed and quality." Mars employs multi-functional teams to smooth the time to market. "We have people with lots of variety of experience, finance, technology, manufacturing, who have lots of different perspectives," explains Boushell.

This focus on cross-functional collaboration also extends globally. Mike Webster, Product Development Manager for sugar products like Starbursts™ and Skittles™ explains that the collaboration goes well beyond the obvious idea of trying to extend the Lucas brand of candies that dominate the Mexican market into the United States: "We have annual meetings across global R&D because so many of our products are shared across geographies.

For example, our annual global meetings include representatives of the three technology groups, United States, Europe, and Australia, but also invited guests from other functions such as finance, supply, and marketing."

Mars associates regularly transfer across global organization boundaries that helps create cross-functional linkages that facilitate collaboration through personal relationships. Webster elaborated, "We'll take a short-term pain of staff loss in one section to give a long-term benefit to the business."

In fact, John Helferich views process discipline and global collaboration as keys to his job. "The entire company is now looking at all business processes and new product development ranks among the top priorities. In my functional role, I have to make sure that strategic connections are established, best practices get transferred and synergies are captured." And so goes the ongoing innovation of innovation.

Putting It All Together

* * *

Mars offers a good example of how a company totally transformed its approach to new product development by shifting from iterative product extensions to consumer focused innovation based on market research and customer insight. The transformation has resulted in some of the industry's most innovative new product introductions.

But as Mars acknowledges, innovative product design does not necessarily guarantee success. The company remains vigilant about ensuring manufacturing and packaging efficiencies and cost savings by leveraging existing technology and processes whenever possible. And by instituting formal stage-gate and portfolio management processes, the company feels it has gained significantly greater management control over product creation enterprise-wide.

Chapter 10

WHIRLPOOL CORPORATION

———— * * * ————

Building a Global Product Creation Capability

In an industry long driven by fierce price-based competition, Whirlpool Corporation's strategic shift to consumer-driven innovation has actually spurred a transformation in the home appliance industry's competitive environment worldwide. The company, in fact, publicly proclaims that "Innovation is Whirlpool Corporation's differentiating strategy; one we believe provides us with a significant competitive advantage."[1]

Whirlpool doesn't just pay lip service to innovation. The company has truly put its money where its mouth is by developing a dynamic product creation capability and fully embedding it within the company's organizational culture enterprise-wide. Today that enterprise literally spans the globe—the corollary to Whirlpool's innovation transformation has been the globalization of its product creation, procurement, and manufacturing processes.

---

[1] www.whirlpool.com.

With annual sales of more than $19 billion, 80,000 employees, and nearly 50 manufacturing and technology research centers worldwide, Whirlpool Corporation has become the world's leading manufacturer and marketer of major home appliances—often referred to as white goods. The company markets Whirlpool®, KitchenAid®, Maytag®, Brastemp®, Bauknecht®, Consul®, Jenn-Air®, and other major brand names to consumers in more than 170 countries. It also serves as a principal supplier for many of the major home appliances that Sears markets under the Kenmore® brand.[2]

That Was Then

* ✳ *

For nearly all of its recent history, the white goods industry competed almost exclusively on cost and quality, to such an extent that yearly product goals at Whirlpool in the early 1990s might have been to take out $1 to $2 per clothes dryer unit with the cost reduction goals set at literally pennies for each of the different functional parts and systems. Such cost-centric goals were common practice across the entire industry. In fact, most lower-end washers and dryers literally sold for about $1 *per pound* retail, and that's for a product with an expected life of 15 to 20 years.

Not surprising, product innovation within the industry had stagnated: the last burst of true product innovation dated back to

---

[2] ibid.

the 1960s following Litton's introduction of the microwave oven. Virtually all white goods companies were led by engineering and manufacturing executives, and their cost-orientation pervaded the entire industry. In addition, most products and components were manufactured domestically. Purchasing, as might be expected, focused on cost and quality, and the buying group often consisted of those who couldn't make the grade in the core functions of engineering and manufacturing.

This Is Now

* ✳ *

Today the white goods industry has undergone a sea change, and it shows. Global players—including Whirlpool, Electrolux, GE, Bosch Siemens™, L.G., and Samsung—operate from major economies in the United States, Europe, and Asia. In like fashion, the industry's supply base has gone global. The increased focus on innovative products has compressed the product model/market life cycle to about three years, while the product useful life in the home remains an impressive 15 years. Stringent cost and quality standards continue to dominate development and manufacturing, but innovation and differentiation wield power in the marketplace.

Retailing giants such as Sears, Carrefour, Sam's Club®, Best Buy™, Lowe's®, and The Home Depot® dominate the sales channel, primarily by competing on price. Such powerful, cost-focused competition keeps margins low and forces manufacturers to seek economies of scale though high volume production. Looking forward, the power of manufacturers in the sales channel depends largely on their rate of innovation.

Strategic Transformation

* * *

In the 1980s, Whirlpool faced a mature appliance market in the United States with domestic sales growth measured in fractions of a percentage point. In other parts of the world, however, trade barriers began to fall, and new free enterprise economies flourished. Under the leadership of David Whitwam, who took the reigns as CEO in 1987, Whirlpool began a strategic transformation that continues to define the company today.

Whitwam, who came to his new role with a strong background in sales and marketing, sensed growing customer frustration with his company's lackluster products, heavily influenced by the company's pervasive engineering and manufacturing mindset. In response, Whitwam crafted a vision for Whirlpool with two mutually sustainable goals:

1. Take the business global

2. Focus on consumer-driven innovation and brands

While these goals seemed daunting at the time—and certainly rife with challenge—most would now agree that Whirlpool's survival demanded such monumental change. Although initially focused on growth, Whitwam's strategic vision served as the underpinning for one of the industries' most robust and comprehensive product creation capabilities. Ultimately, that capability not only positioned Whirlpool as the leader in the U.S. appliance industry, but also as the largest white goods producer in the world.

Global Expansion

* ☀ *

Whirlpool had long operated an international division with joint ventures in Brazil and Mexico and export sales to Taiwan, Hong Kong, and Australia. Whitwam sought a much more global footprint and began expanding into other markets. For example, in 1987, Whirlpool Corporation and Sundaram-Clayton of India formed TVS Whirlpool Limited to make compact clothes washers for the Indian market.

In 1989, Whirlpool made a major move into Europe through a joint venture with N.V. Philips of the Netherlands to manufacture and market appliances in Europe. Two years later Whirlpool acquired Philips' share and became the sole owner of the business. Combined with other global and domestic acquisitions and expansions, Whirlpool entered the 1990s with revenues exceeding $6 billion.

The acquisition of the Philips white goods business proved critical to Whirlpool's globalization strategy. It gave the traditional, midwestern manufacturer a much more global management perspective and expanded operations in Asia, South Africa, South America, and Europe. The acquisition also underscored the necessity for another pivotal transformation within Whirlpool. Although Whitwam had organized the business on a regional basis to ensure local market focus and accountability, he envisioned a global new product creation organization to capture economies of scale and to support synergies within the worldwide enterprise.

Focus on Innovation

* * *

In the wake of the Phillips acquisition, Whirlpool faced the significant tasks of integrating and globalizing the combined companies and, at the same time, reinventing its product creation process. Two key changes initiated by Whitwam enabled Whirlpool to accomplish both tasks.

First, Whitwam created what he called "Global Challenge Teams," which compelled executives and key players from the different geographic regions to leverage their combined capabilities collaboratively. Second, he hired Ron Kerber, a seasoned McDonnell Douglas executive with experience in engineering, marketing, strategic planning, and procurement. Kerber assumed the newly created position of executive vice president and chief technical officer with the broad mandate to globalize product creation and procurement. Kerber now wryly admits that at the time neither he nor Whitwam "really knew what that meant."

Nonetheless, he set to work by creating two separate teams composed of and led by internal staff. He avoided using outside consultants to ensure clear ownership by the organization. Kerber charged the two teams respectively with:

1. Proposing a new organizational structure for managing technology and product creation

2. Developing a product creation strategy and process

Launched with CEO Whitwam's personal endorsement, the teams proceeded in parallel with periodic updates to the chairman and executive committee.

Organizational structure. The organizational structure study team, composed of vice presidents from the corporate office and the regional businesses, moved forward quickly. Their efforts resulted in the creation of a global chief technology officer (CTO) group that eventually encompassed:

- Corporate research and development

- Global product development

- Global procurement

- Advanced products concepts

- Strategy and analysis

- Global business units for compressors, microwave ovens, and air treatment

Product creation strategy and process. To ensure a high-level, multi-disciplinary perspective, the product creation strategy and process team included all regional marketing, sales, and manufacturing vice presidents as well as the corporate vice presidents of research and development and procurement. The team began by diligently benchmarking product creation at other leading corporations, including Ford, Hewlett-Packard, 3M, and Procter & Gamble. To reach their desired end state—a clearly articulated product creation strategy and process—the team also examined Whirlpool's current processes and assumptions to identify what was right with them, what was wrong, and what should change.

Their hard work paid off. Guided by the strong analytical and strategic skills of team member, Dan McNicholl (a McDonnell Douglas veteran and the newly appointed VP of strategic planning and analysis for the CTO organization), the team's efforts had two primary tangible outcomes. First, they articulated a comprehensive new Whirlpool product creation approach, including clearly defined expectations for project structure, management, and multi-disciplinary functional participation. A component of the broader process, and absolutely critical to its implementation, adoption, and success was the C2C—consumer to consumer—process (described later), which would serve as the definitive guide for all new product creation projects going forward.

With process and documentation in hand, the team next faced the challenge of a company-wide roll-out. The team conducted training sessions starting with the executive committee (the CEO and all EVPs), then senior management, and finally leadership in each region. Notably, Whitwam personally kicked off each roll-out meeting with an explicit mandate: he would not approve capital for any new product creation project that failed to adopt the new process. Not surprising, his endorsement spurred widespread acceptance.

Of the process and its roll-out, Jeff Fettig, then Vice President of KitchenAid Marketing and now Whirlpool's CEO, said at the time, "The synergy created by the new process and the CTO organization that enabled it are remarkable. This roll-out takes Whirlpool to a new level and raises the bar for all corporate planning."

Concept to Consumer: Whirlpool Corporation's C2C Process

✳ ✳ ✳

The C2C process—graphically depicted in Figure 10.1—serves as the centerpiece of Whirlpool's approach to product creation.

C2C, a classic stage-gate project management process (see Chapter 4), has been vital to effective product creation at Whirlpool. Equally critical is the company's overall strategy, approach, and culture. For example, the team's recommendations addressed the need for a better understanding of customer needs,

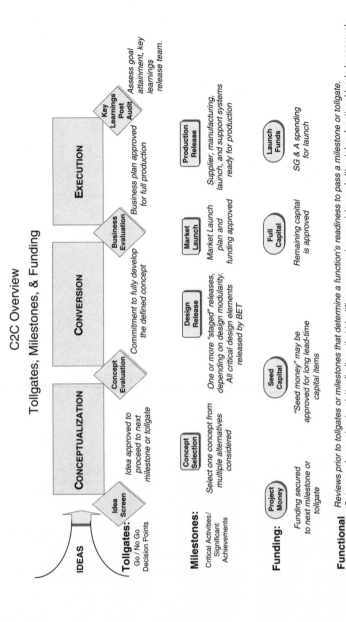

Figure 10.1 The Whirlpool Product Creation Stage-Gate Process. Source: Whirlpool Corporation

the role of technology demonstrators, and multifunctional involvement in product creation.

The team also focused on developing and maintaining a continually evolving portfolio of technology efforts, development projects, and products in the marketplace. Whirlpool comprehensively manages its product investment portfolio by brand, product, and model. To do this, the company relies heavily on developing global product platforms and the use of block upgrades for these platforms. Although other companies use such techniques, Whirlpool's integrated management discipline sets the company apart.

Figure 10.2, an explanatory graphic used at Whirlpool, shows how technology roadmaps must link explicitly to the portfolio of new product creation projects and ultimately to the subsequent upgrades or changes demanded by the product portfolio. Although the arrows suggest a forward flow, the reverse proves equally critical. If marketing, sales, or even a channel partner identifies the need for a specific product upgrade or new product, the process traces back as necessary to identify the technology development that will be required.

Although the C2C stage-gate and the portfolio management processes serve as the backbone for new product creation at Whirlpool, three other aspects of Whirlpool's approach to corporate product creation deserve further exploration:

- Highly disciplined project management

- Consumer and brand focus

- Global integration

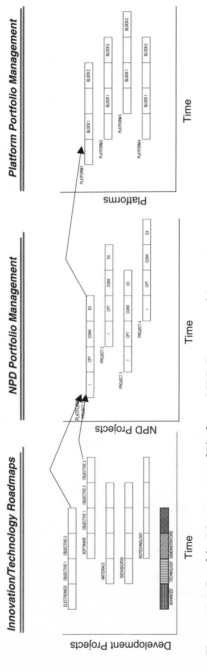

Figure 10.2 Portfolio Management of Platforms, NPD Projects, and Innovations

Highly Disciplined Project Management

* ✳ *

Whirlpool employs an extensive set of tools and techniques to ensure highly disciplined management of all technology projects and development phases. Starting with strong executive sponsorship and committed leadership, the company uses formal goal setting, reviews supported by "red teams", and comprehensive human resource practices that include talent pool management, promotion and reward, and training. It also seeks to continually improve its approach to product creation as conditions and priorities change.

Executive sponsorship. For every new product creation project, Whirlpool names, at the outset, an executive decision-maker and sponsor (usually a vice president), who serves as the single focal point for decision making. Executive sponsorship establishes clear authority and well-defined lines of accountability and responsibility throughout the project. It also helps to ensure that products remain true to the consumer positioning of each brand.

Project leadership. While executive sponsors set the vision for projects, remove obstacles, and support their teams, the most critical task is selecting the right project leader to assume accountability for all project deliverables. The executive sponsor ensures that the leader receives the authority and appropriate resources to achieve the project goals and also provides ongoing support and coaching throughout the process.

Formalized project expectations. The clear list of expectations that Whirlpool sets for product creation management drives project execution and increases the chances for success. Tom Egan, North American Vice President of Product Delivery, explains, "C2C puts discipline into the project. Without it there would be chaos. Everyone understands that there are formal checkpoints that must be cleared before a project can go forward. The decision making and expectations are very clear. There can be debate, but these elements are not up for discussion."

Annual product reviews. Whirlpool's executive committee and senior management conduct formal annual reviews of the company's platform strategies, technology roadmaps, and product creation projects. The executives assess readiness, examine discrepancies in the product portfolios, and look for new opportunities. This discipline not only compels management to revisit and reassess strategies, it also provides a forum to examine in detail the total product investment strategy. Because of their timing relative to the company's business planning process, annual product reviews can result in changes in direction and brand extension opportunities for the following year's business commitment as needed.

Red teams. Whirlpool uses red teams to provide critical and timely analysis of a product creation project throughout its execution. The team includes a diverse group of experienced managers and specialty experts from engineering, procurement, manufacturing, finance, marketing, sales, and service who are expected to offer independent, constructive assessments—both positive and negative—of any aspect of the project including its execution, viability, and readiness.

Human resources development. Whirlpool places high priority on human resource development and looks to product creation projects as a fertile source of and training ground for future company leaders. Before it strengthened its human resource processes, Whirlpool managers reluctantly joined product creation projects fearing being taken out of their functional homes and possibly missing normal promotions and career progression. Now, Whirlpool carefully targets successful product creation project leaders for future key positions recognizing the value of leveraging their demonstrated experience and talents.

Continuous process improvement. Since its launch in 1992, the C2C process has had two major refreshes or course corrections to address gaps and to focus the process appropriately. The first refresh brought the process back closer to its original intent. Over time, C2C had become more bureaucratic and focused on engineering management and process detail than on customer and business benefits. The revision addressed these shortcomings and restored the process to its original intent.

More recently, a second refresh of C2C enhanced the ideation or concept phase of the process. Now known at Whirlpool as the "fuzzy front end," this enhanced process element brings the consumer point of view and brand promise even more directly and meaningfully into the very early stages of product creation.

Whirlpool has a corporate product creation process owner, typically a global product area vice president responsible for the health of the process, minor upgrades, and company-wide training. Such executive accountability has proven very helpful in keeping all players and the process itself on course.

Consumer and Brand Focus

* * *

Whirlpool's enhancement of the fuzzy front end helps keep product creation focused on customer needs and in line with the brand image. It sets the expectation that brand concepts will be translated in product lines and helps guide how new products will link to the different elements of strategy. Tom Egan maintains, "There's a very clear linkage to longer-term plans, the brand promise, how that translates into products, how the brand serves as the guiding light, and how that works through brand and channel performance."

The process for focusing more intensely on consumers and brands has grown far more disciplined and thorough than originally conceived. Whirlpool spent considerable time, money, and effort to instill the concept of consumer-focused innovation into the organization. This mindset and staying true to the brand promise now form the essential starting point for any product creation project.

Many lenses for understanding customers. To forecast and understand customer needs, Whirlpool looks ahead 10 to 15 years to try to determine future consumer demographics and lifestyle wants and needs. They use a wide variety of lenses to capture the voice of the customer including focus groups, direct observation, and ethnographic research. The company has also innovated or enhanced a number of techniques including process mapping, innovation cells, "concept cars," and usability laboratories.

An often cited example of how Whirlpool gets into the mind of the customer may sound mundane, but actually proves quite powerful. Product creation team members ride along on a Whirlpool service truck on its regular route to listen to complaints and to observe how customers use the company's products in the home. Over the years these observations have provided great insight into the uses—and abuses—of Whirlpool products. Tales abound of unbelievably smelly kitchens, attempts to "dry" cats in a microwave, and "cooking" fish in a dishwasher. Odd as some of this may seem, such on-site research adds significant depth to Whirlpool Corporation's customer understanding and knowledge base.

The green line. Whirlpool uses the green line concept, to support its customer focused process. Any aspect of a product that the consumer can "see" or perceive and value lies in front of the green line; everything not "visible" to the consumer lies behind. The green line concept helps designers build-in as much product differentiation as possible in front of the green line to support the brand and customer preferences. Behind the line, they maintain common platforms and components to keep production costs as low as possible.

For example, in a refrigerator, the cabinet door, industrial design, shelves, bins, and control system lie in front of the green line. The cooling system and insulation are behind it. Sometimes the green line shifts. When Whirlpool promoted the "Catalyst®️ Wash System" similar to the "Intel Inside" promotion, they brought a feature from behind the green line and positioned it in front as a consumer benefit.

Global Integration

* * *

In the years since it first began deploying its revamped product creation process, Whirlpool has developed and embedded within its global enterprise the skills and capabilities required to discover, develop, and rapidly bring to market true innovation. Such innovations are part of Whirlpool's strategy of building customer loyalty by applying a deeper, more fundamental understanding of customers' needs, desires and lifestyles.

Globalization of product creation. Today, cross-regional teams of Whirlpool product creation engineers consolidated in a single global organization collaborate on innovation initiatives for both regional and global distribution. The Whirlpool® Duet® clothes washer and dryer pair launched in the United States in 2002, serves as a prime example of how such collaboration can result in a highly successful product.

The product integrates European front-load washer technology with a design that appeals to U.S. consumers. Developed by engineers in Schorndorf, Germany, based on U.S. market functional requirements, manufactured in Germany for sale in the United States, and paired with a dryer made in Marion, Ohio, the Duet system rates as one of the company's most successful new products in years—a perfect example of leveraging a global technology organization. Other examples include a frost-free refrigerator designed in Brazil, first sold in India, and ultimately in China and Brazil and a new over-the-range microwave oven and hood designed in Sweden, manufactured in China, and sold in the United States.

With the globalization of product creation, the lenses Whirlpool uses to understand consumers don't reveal the same answer in every region. Knowing what to look for or observe in one region, however, can lead to the discovery of new customer benefits in others. Some benefits that have migrated globally include electronic controls and consumer interface to products, frost-free technology, refrigerator interior design, dishwasher design, clothes washer process design, and the company's quality process design.

Integration of product creation and procurement. Whirlpool organized procurement globally at the same time that it moved to a global process for product creation. The next big move with much bigger impact and strategic benefit was the globalization of its product development organization. To demonstrate the benefits of this new combined global organization, "quick win" teams in each product area from the engineering and procurement groups set out to find redundancies, common components, and other cost and quality opportunities. They focused their initial efforts behind the "green line" so that cost or quality improvements could be made, but the perceived product would remain the same. As a result, the teams not only cut cost by several percentage points, but also turned those who doubted the value of globalization into believers.

Today, the company seeks to leverage global designs and pursues global sourcing strategies routinely. For example, buyers dedicated to supporting the same product lines in different regions regularly benchmark pricing of key components. The exercise not only enables better price negotiations, but also often uncovers opportunities for component standardization across the globe. Whirlpool views the integration of product creation and procurement on a global basis as a key enabler of its strategy. In fact, Whirlpool appears to be the only large durable goods manufacture to have integrated these functions.

Prior to development of C2C, procurement had no formal voice in the product creation process. Now, with more engagement and more accountability combined with the right product creation structure and decision points, procurement contributes meaningfully alongside the other disciplines to an integrated development process.

Leveraging supplier innovation. Whirlpool seeks supplier involvement in its innovation process by focusing on what Jerry Weinstein, Senior Vice President of Global Procurement, calls "targeted innovation." He says, "We don't just ask suppliers to give us their best ideas, their ideas have to fit into our brand space." To that end, Whirlpool trains suppliers in its innovation process and understanding of the brand promise.

Whirlpool also fosters an environment that encourages suppliers to bring innovation forward. The company sets aside time during its periodic supplier council meetings to give suppliers the opportunity to present their ideas to high-level executive officers rather than forcing them to run the multi-layered buyer/engineer gauntlet to bring new ideas to the fore.

Weinstein adds, "We would not limit our search for innovation to our current supply base. We might go to another industry where a supplier has solved a technological problem. That means our procurement people will have to know those supply bases as well."

Cultural integration. Key to Whirlpool's overall global integration has been the development of a common language and management approach. The Whirlpool product creation process, the C2C component in particular, has given the company a common vocabulary that crosses all geographic regions worldwide. According to Michael Thieneman, Whirlpool Corporation's Chief Technology Officer, "Without it there would be a Tower of Babble. More important, with customer loyalty as the centerpiece of our architecture, the product creation process has been extended and is enabling our current strategy."

Marise Kumar, North American Vice President of Innovation and Core Competencies, confirms the value of a common language in describing a recent organizational change that required teams from very different cultures to establish good working relationships immediately. "C2C helped both groups communicate because they were using a common language," she says. "They used the process not only for getting the product out, but also for forecasting, order processing, and distribution. Perhaps you can't transfer tribal knowledge, but C2C got us to the stage where we could supply consistently from one region to the other for the coming season. It was a facilitator and enabler."

An Organizational Capability

* * *

While Whirlpool would have to wait for its new product creation process to show significant advantages, incremental benefits came much quicker. For example, projects were completed on schedule and within budget and cycle times decreased significantly. In terms of new product launches, the results over time have proven dramatic, as Figure 10.3 shows.

The decade-long product creation transformation has moved Whirlpool from an engineering/manufacturing focused company to a consumer-driven global enterprise with a dramatically

---

Results of Product Creation Process Redesign

Polara® refrigerated range

A combination refrigerator/oven that keeps a dish cold until it is time to bake or heat. It allows working people to prepare a meal before leaving for work and have a hot entrée ready when they come home.

Duet® Laundry Pair

An upscale ergonomically and environmentally friendly front load washer and dryer. Resting on its custom pedestal it loads at waist height, washes clothes better and with less water, and removes more water in the spin cycle for more efficient drying.

Calypso® washing machine

A top-load washing machine without an agitator that washes clothes more gently than, yet as well as, a front load washing machine using less water.

Laundry rooms

Integrated work centers that enable all types of laundry tasks from traditional washers to "drying drawers" for air drying delicates in an organized way.

Gladiator® Garage Works

Rugged steel diamond plated devices that help organize the typical suburban garage into a neat, functional work center.

Personal Valet® Clothes Revitalizing System

A clothing care device that de-wrinkles and refreshes by reducing offensive odors from dry-cleanable clothing. Allows multiple uses of a garment before it must go for professional dry cleaning.

Internet-enabled oven

An oven that can be turned on or off remotely over the internet.

Figure 10.3 Recent Whirlpool Innovations

energized innovation machine. According to Michael Thieneman, Whirlpool launched 71 "mega" or major projects in 2004. "A decade before, you could have counted them on one hand," he says. Thieneman also says that the velocity of product cycle time continues to improve. For example, in the three years ending in 2004, the company launched 24 industry innovations. In the year and a half from 2005 to midyear 2006, projections call for a total of 22 industry innovation launches—nearly the same number in half the time.

Such dramatic improvement cannot be attributed, much less accomplished, by a single function or in isolation by a few. It requires an integrated multi-disciplinary approach, disciplined management, and most of all a culture that values and nurtures innovation. Whirlpool management would say the product creation process has accomplished all that and is firmly embedded as a true organization-wide capability. In fact, product creation has inspired the redesign of other core business processes at Whirlpool.

Jeff Fettig, Whirlpool Corporation's current CEO, says "The product creation process is embedded in the company fiber. It is one of four key foundation processes by which we mange the company and measure our success. It has given us a common language, enabled globalization, and driven unprecedented product innovation."

Chapter 11

UNITED TECHNOLOGIES CORPORATION

———— * ✳ * ————

Discipline and Balance in Innovation

The $43 billion industrial powerhouse, United Technologies Corporation (UTC) has long remained low on the horizon to the average consumer. True, most anyone who has ridden in an elevator has seen the name Otis. Likewise, many would recognize Carrier as the worldwide market leader in heating, ventilation, and air conditioning systems for residential, commercial, industrial, and transportation customers. But most wouldn't know that Carrier is a UTC brand. And even though Pratt & Whitney, Sikorsky, and Hamilton Sundstrand are critical suppliers of aircraft engines and major aerospace systems for most of the world's commercial and military aircraft industry, few outside of those industries even know the names, much less the fact that they are UTC business units.

So what is actually in a name? The average person on the street may not know even how to spell UTC, but savvy investors surely do. The business and investment communities have definitely taken note of the 975 percent shareholder return that UTC produced between 1993 and 2004 under the leadership of chief executive George David. Named 2005 CEO of the Year by *Chief Executive Magazine*, David—and by extension UTC—has clearly stepped out of the shadow of Connecticut neighbor Jack Welch, former CEO of GE. Widely recognized as one of America's best run companies and a key competitor to UTC in aircraft engines, GE delivered a "mere" 561 percent return over the same 11 years, ironically a period that started when Welch was named the magazine's 1993 CEO of the Year.

David attributes UTC's sustained performance to "operating discipline leading to significant cost reductions and game changing products."[1] To maintain such discipline, he also supports a highly decentralized organization model and a culture of accountability throughout the enterprise. In an interview announcing his selection as CEO of the Year, David says, "I'm a tremendous believer in profit-center management. You use the discipline of the market to cause companies to perform."

According to John Cassidy, retired Senior Vice President of Science and Technology for UTC, that decentralized model presents its own set of challenges as well, "Our culture reflects a set of fiercely independent tribes…who regularly demonstrate their independence with their war paint." Despite the challenge of orchestrating a 450-person central research center with thousands of technology experts scattered throughout a vast decentralized entity, Cassidy applauds David's constancy of purpose. "Some CEOs move back and forth," says Cassidy. "David is boringly

---

[1] United Technologies Corporation 2004 Annual Report, Chairman's Letter to the Shareholders.

consistent. His principles and emphasis have remained the same for a decade."

That is the story this chapter tells. How those principles and emphasis—operating discipline, cost reductions, game changing innovation, decentralization, and a culture of accountability—have sustained UTC as an industrial powerhouse over the long haul.

Multi-Industry, Multinational

*** *** ***

UTC eschews the term "conglomerate" in favor of the more descriptive "multi-industry, multinational," which George David defends, not surprisingly, from a financial market perspective, "Investors hate bad news. The way you get no bad news is by having the company be very far-flung both geographically and product-wise."[2] And far-flung it is. About 60 percent of UTC revenues derive from outside the United States. That number runs as high as 80 percent for world market leader Otis. And despite a history anchored in the aerospace industry, 64 percent of UTC revenues now come from commercial and industrial markets—Otis, Carrier, and UTC Fire & Security primarily—with the military and commercial aerospace businesses accounting

---

[2] Holstein, William J, "George David Steps Out," *Chief Executive Magazine*, Vol. 208, May 2005.

for 18 percent each. UTC also has a healthy mix of original equipment manufacturing (at 58 percent) and aftermarket products and services (42 percent).

That robust mix has emerged through acquisitions and divestitures involving companies dating back to the beginnings of the industrial revolution. The company's name, however, dates back only to 1975 when United Aircraft Company, a product of a 1934 federally mandated divestiture (which also created Boeing and United Airlines®) renamed itself United Technologies Corporation. The company's new moniker captured its existing broad product base as well as its future ambitions. In line with those ambitions, the company quickly added two major acquisitions outside of the aerospace industry.

First, in 1976, UTC acquired Otis Elevators, a company with a history that goes back to 1853. Then two years later, UTC acquired Carrier Corporation, originally founded in 1915. These commercial pioneers—one had invented elevators and the other air conditioning—became part of UTC's multi-industry, multinational venture that included Pratt & Whitney, Hamilton Standard, and Sikorsky, each of which had a comparable history of pioneering, albeit in different industries, dating back as far as the 1920s. Like their older commercial siblings, these aircraft—and later aerospace—businesses had been industry innovators with breakthroughs ranging from the invention of the helicopter in 1939 to the Apollo space suits in which astronauts bounded across the surface of the moon in 1969. (See Figure 11.1 for one of the latest innovations from UTC's Sikorsky business unit.)

In addition to dozens of minor acquisitions over the ensuing years, UTC's Hamilton Standard Division merged with Sundstrand

Figure 11.1 Sikorsky's S-92. The Sikorsky S-92 raised the standard in helicopter safety, comfort and reliability

Corporation in 1999 and UTC acquired Chubb in 2003 to create the current set of seven operating companies. Otis, Carrier, Power, and Fire & Security, comprise the current set of commercial businesses while Pratt & Whitney, Hamilton Sundstrand, and Sikorsky make up the aerospace side.

A Culture of Discipline: ACE and Passport

* * *

The recent success of UTC largely derives from a culture of disciplined, continuous improvement that traces back to an Otis co-branded joint venture with Matsushita in the late 1980s. Matsushita, consistent with the Japanese fanaticism with quality, insisted that Otis attain the same standards of excellence demanded in the home market of Japan. According to company lore, as passed on by John Cassidy, the president of Matsushita essentially told Otis, "You are an embarrassment to us."

The Ito influence. That inauspicious beginning led to a propitious meeting between George David and Yuzuru Ito, a key Matsushita executive and quality guru. The Ito approach to quality applied simpler tools than those advocated by Six Sigma proponents, but ones which enabled active involvement of everyone, including front-line employees. Ito retired from Matsushita, but continued to serve as a consultant to UTC, reporting directly to David, until Ito's death in 2005.

During the 1990s, Pratt & Whitney began to apply the "lean concepts" employed at Toyota and popularized by James P. Womack and Daniel T. Jones[3] based on their groundbreaking research comparing Japanese and Western automotive manufacturers. "In late 1995, some Pratt people put the two philosophies together and named it Achieving Competitive Advantage (ACE)," according to

---

[3] James P. Womack and Daniel T. Jones, *Lean Thinking*, Simon & Schuster, 1996.

Ralph T. Wood, former Director ACE, in UTC's Quality group. "In 1998, Ito suggested we create a quality university to teach ACE across the entire enterprise." The new training institute became known as Ito University in honor of the man who helped instill what David refers to as "a powerful repetitive discipline" in the UTC culture. David attributes at least half of the shareholder gains achieved in his tenure to disciplined application of ACE.

A renewed commitment to process improvement. CEO George David's emphasis on disciplined results also stimulated the recent renewal of UTC's product development process. In 1995, noting that UTC outspent its competitors on research and development without producing an obvious superior return on the investment, David issued a challenge to his entire engineering community to make marked and rapid improvements. The initial efforts generated a lot of work but few results. "The problem was viewed in terms of engineering productivity," explains Wood. "We developed a diagnostic tool and documented a huge number of 'best practices' and tried to come up with metrics to see if gaps were being closed. There was no focus at all, and it fell apart within a year."

David then charged the quality and technology groups to merge their thinking relating to "engineering productivity," which led to a focus on the "cost of quality." UTC calculated a $1 billion price tag for three critical cost categories: warranty payments; scrap, rework and repair in manufacturing; and engineering rework. "We were a $20 billion company at the time and the cost of poor quality was equal to the total corporate profit," notes Wood. This observation obviously captured the attention of the business unit managers, the critical sponsors needed to drive change in the decentralized company.

Leveraging the momentum that was growing at the business unit level, John Cassidy, then Senior Vice President of Technology and Science, launched a stage-gate process to put better business discipline into UTC's research and development spending—a total that now exceeds $2.5 billion per year. According to Cassidy, "In the badness of the past, we would declare a project complete when time and money ran out, not necessarily when the product was fully developed, tested, and validated."

In characteristically disciplined style, Cassidy benchmarked his newly energized effort against the auto industry, Motorola, Xerox, and GE—"the usual suspects"—according to Cassidy. The new system, called "Passport," grew from the quality culture as a disciplined way to drive decisions. Explains Cassidy, "Passport is about decision making and risk reduction. We made sure that it is multidisciplinary in order to engage all the functional people, to get everyone to buy in." The Passport process instills the discipline and courage needed to cancel projects that don't meet required hurdles, including financial returns. It puts the same discipline into the managerial process and decision making. "The relevant business unit president must be involved," explains Cassidy. "We cancel the meeting if key members can't attend."

Given UTC's decentralized culture, Cassidy feels his focus on the discipline of Passport helped avoid potential backlash from the decentralized operating companies about "centralized bureaucracy." For example, even though the Passport discipline pervades the company it is adapted to the specific needs of the function or business unit. Pratt & Whitney and Otis have the autonomy and flexibility to describe their specific stages and gates differently, using jargon more common to their respective industries.

Passport is much more than an engineering discipline: UTC had always produced well-engineered, highly reliable products. A company doesn't become the dominant player in safety critical products like jet engines, elevators, and helicopters without superior technical competency. Passport's distinction lies in its embedded concept of "earned value" intended to bring greater discipline to the development process. Art Lucas, Senior Vice President Engineering, Pratt & Whitney, explains, "By earned value, we mean that the company measures each development project against pre-declared commitments relating to scheduling and cost."

Lucas continues, "Quarterly, I report on what engineering is doing directly to the chairman. He tracks 'earned value' and progress against our 'job ticket.' The job ticket tracks the key product performance parameters such as weight, efficiency, reliability, and observability—key to military stealth applications. We sign up with each customer for what we'll deliver in each category. We need the job ticket because I might be meeting the schedule and budget, but not delivering the right product," says Lucas.

Pratt & Whitney Canada's design of the new PW6000 jet engines, which will target smaller corporate jets, demonstrates the power of applying multifunctional discipline. By assembling a global team of technical experts from the business unit as well as from the supplier base, Pratt & Whitney says it will revolutionize the engine design and manufacturing process, with a new design that contains half as many parts and requires eight hours instead of eight days to manufacture.

The Voice of the Industrial Customer

* * *

UTC's improved process discipline and its greater emphasis on maintaining closer ties to customer requirements has recently helped the company focus more intently on capturing "the voice of the customer." The concept serves UTC's product development process well, despite the company's industrial as opposed to consumer focus. At Otis, for example, product managers for each of the major product lines serve as the key contact points for collecting field information—such as input from the sales representatives and anecdotes from service/repair personnel—and formal customer survey results as well. The product managers ensure that engineering not only "hears" but "listens" to the many different customer voices. For example, an Otis product manager might insist that technical people put themselves in the shoes of passengers by riding elevators in actual service, speak to building owners about maintenance, discuss with general contractors such issues as initial cost and installation, capture the architects view of aesthetic expectations, and confer with "elevator consultants" about performance requirements for servicing a given passenger flow pattern.

"I tell the product managers that their job is to be the engineers' biggest 'pain in the butt' before the first tollgate to make sure they set the proper 'terms of reference' for the development project" says Sandy Diehl, Senior Vice President of Product Strategy at Otis. "After that point, the product manager's job is to make the necessary tradeoffs that will help the engineers look great by meeting the remaining tollgates on time and on budget."

To understand how UTC uses the voice of the customer to maintain a proper portfolio of products, consider the range of efforts Otis expends to maintain its leading market share. The company's innovation projects range from "new to the world" products to fundamental re-conceptualizations of existing products to ongoing renewals that improve the customer experience. At the simple end of the continuum lies an effort the company undertook to simplify the elevator installation process for contractors. Otis deconstructed and analyzed the entire installation process to gain insight into the difficulties of physically transporting parts, which at that time arrived at the installation site in a range of differently sized packing crates. Based on the result of the analysis, Otis simplified installation significantly by reorganizing parts into a variety of "kits" and by developing a patented hand truck that would allow the installer to move kits through the doors of any building without having to break down packaging.

At the other extreme of the product development continuum, Otis pursues "radical innovation"[4] as exemplified by its Odyssey elevator project in the late 1990s. The challenge? Create an elevator system for the "mile-high building" first envisioned by Frank Lloyd Wright in the mid-1950s. Although Wright at the time had available the materials and technology to support the construction of his mile-high vision, a building of such height was totally impractical with existing elevator technology. The weight of the elevator cable alone imposed practical limits on the height of any single shaft and, even without that limit, efficiently moving

---

[4] The description of the Odyssey project is drawn from *Radical Innovation: How Mature Companies Can Outsmart Upstarts*, Leifer et al, Harvard Business School Press, Boston 2000.

people among hundreds of stories would require an inordinate number of separate shafts, each consuming valuable floor space throughout the full height of the building.

The Odyssey elevator system concept developed by Otis offers a solution to the problem with a design that allows an elevator car to change shafts by moving horizontally as well as vertically. Such flexibility means that more cars can effectively operate with fewer shafts. While the Odyssey concept resonated with Hong Kong developers at the time, the financial crisis of the late 1990s dubbed the "Asian Flu" wiped several super-tall building projects off the drawing boards and forced Otis to shelve the project.

Although Odyssey may have been ahead of its time, Otis leveraged many of its key concepts in developing the company's more mainstream Gen2™ elevator design (see Figure 11.2). The Gen2 elevator uses a flat, polyurethane-coated steel belt to lift the elevator car eliminating the steel cables that had been a mainstay of elevator design for more than a century. Inspired by the self-contained concept of the Odyssey and enabled by both the reduced weight of the steel belts, the Gen2 design resulted in a 70 percent reduction in the size of the equipment that lifts the elevator, which does not require a separate machine room. The Gen2 also supports UTC's goal to improve environmental sustainability: it eliminates the need for lubricants and can reduce energy consumption by up to 70 percent in some applications.

UTC excels at portfolio management despite the challenges of product lines comprised of highly engineered equipment that can take years to develop and have productive lives measured in decades. Part of the company's portfolio management success is not only knowing when to make big bets, but also how to salvage

Figure 11.2 Otis' Gen2 Elevator design employs advanced technology capabilities to save space and reduce energy requirements

intellectual capital even when a bet, such as the Odyssey elevator, does not pay off as expected.

Just as important to UTC's portfolio management challenges is to take a long-term view. Consider Pratt & Whitney, which has only four or five engine programs in development and fewer than a dozen in active duty at any given time. Art Lucas explains, "We can't do an upgrade to improve engine performance. If we need a new engine we have to develop it from scratch. Once you've determined an engine's thermodynamic cycle, you can't tweak the design. It's specifically tuned to an aircraft program." With so much dependent on key, upfront decisions, a long-range point of view becomes critical. "We look 20 years out using lots of customer interaction and feedback," adds Lucas.

Investing in People

* * *

UTC pays more than traditional lip service to investing in people, truly this company's most important asset. The company launched its Employee Scholar Program in 1996, and according to George David,[5] "We have the most robust employee education program in the world. We give paid time away from work on the basis that time is a scarce resource in the modern life—not money. We pay all the tuition and fees. Importantly, we put no limitations on course work. None. You can be an engineer and study law, or theology, or English, or whatever you choose."

UTC also rewards program graduates with up to $10,000 in stock. Since inception, participants have earned 15,800 degrees, and current enrollment totals 13,500. To date, UTC has invested nearly half a billion dollars in the program which is also extended

---

[5] Holstein, William J., "UTC's Global Primer," *Chief Executive Magazine*, Volume 203, November 2004.

to displaced employees for up to four years after a job loss. That's commitment.

Home schooling, too. Not surprisingly, the company leverages its home-grown expertise through in-house technical training programs as well. "We have Fellows, the best of the best, who conduct training sessions in addition to doing their jobs," explains Art Lucas. "That effort totals in excess of 50,000 hours and more than 400 classes a year. We're not trying to teach college-type courses, but rather courses centered on our design process, the tools we use to execute, the design criteria they must design against, or the material specifications that might be different."

Gregory L. Powers serves as Vice President of Engineering at Carrier, UTC's largest operating company at $10.6 billion in revenues. Of his technical "people assets," comprised of 2000 engineers spread out across 72 locations, Powers says, "We're thinking about our talent from the perspective of core competencies. We know what competencies we need and now we are building our reference library. I'm the librarian." Powers elaborates on how this "lending library" taps into the global talent pool, "We launched our first virtual product team in the fall of 2004. Designed in France with support from the United States, the product will be manufactured in four different countries with identical production facilities."

Tes Aklilu, Vice President of Quality for UTC, elaborates on the significance of the global coordination model, "Historically each of the four factories would have developed their own production processes, but now we have institutionalized a 'mother plant' concept. Today when you go to the facilities manufacturing this product in the United States, China, and

Brazil you will see that they are identical to the mother plant in France—the fixtures, the tooling, and even the production planning process."

Synergistic, Breakthrough Innovation

✳ ✳ ✳

Despite an emphatic commitment to decentralization, UTC operating units all benefit from the centralized technical resource, United Technologies Research Center (UTRC). The research center splits its resources between a Components Department and a Systems Department. According to the UTRC website, the Components Department focuses on "innovation in high performance devices" through the use of "materials selection, structural analysis, fluid mechanics, chemical and kinetic modeling, computational fluid dynamics, and thermal management."[6]

As the name clearly implies, the Systems Department applies systems engineering to understand the interactions among components and subsystems of products, services, operations and logistics. UTRC nurtures deep systems competencies in "energy systems, HVAC, chemical processes, combustion dynamics, noise and vibration, dynamic systems and controls, diagnostics and prognostics, operations research, and computational methods and systems."[7]

---

[6] United Technologies Research Center website at http://www.utrc.utc.com.
[7] Ibid.

Or as George Powers of Carrier explains without the technical jargon, "We send the stuff we're not sure of down to UTRC in East Hartford (Connecticut). We give them funding and ask them to work on it. Will it work and if so how? They have their own stage-gate process. Their final gate is our starting point."

UTRC also seeks to create synergy in the white spaces between the existing businesses. In 2002 John Cassidy and Carl Nett, Director of UTRC, hosted a two-day brainstorming session with the top technical people from across the company to seek out fresh ideas for organic growth.[8] The top idea emerging from the session combined the cooling and heating expertise of Carrier with the power generation competency at Pratt & Whitney: a breakthrough system for converting industrial waste heat to power. Carrier uses generators to move air across heat exchangers to produce cool air. Pratt & Whitney uses turbines in industrial applications to generate power. UTRC had the component and systems expertise to help the business unit collaborators model the product using existing components but with the basic flow model reversed.

The UTC Power business installed the first commercial applications of this breakthrough thinking in the form of a combined cooling, heating, and power solution in 2004 at a grocery store, a library/museum, and a public university to demonstrate the huge potential of the technology. UTC Power also leads in the development of clean, environmentally friendly fuel cells from which the only "exhaust" is water vapor. Currently its fuel cells

---

[8] *"A Practical Guide to Social Networks"* by Rob Cross, Jeanne Liedtka, and Leigh Weiss, *Harvard Business Review*, March 2005.

power applications ranging from commercial buses to the space shuttle, and it has development partnerships with organizations ranging from BMW to the U.S. Department of Energy to produce fuel cells for automobiles.

Being the Best, for the Long Haul

* ※ *

Without a doubt, a company that has a history of innovation dating back to the mid 1800s and that also invests in leading edge technology for aerospace and fuel cells takes a long-term view. UTC also aspires to be the very best in everything it does. As George David makes clear, "You make a profit because people pay you more than it costs you to do things. The way you get that is by being very good at what you do. And being very good at what you do to me is much broader than simply current period costs versus previous period revenue. It's the investment in future products. It's the investment in the quality of the workforce. It's the investment in environmentally compliant work."[9]

Despite an enviable record of steady profit growth, David clearly values the long term over the short. "Good organizations win in the world," he asserts. "They may not win in the next quarter or two—but in the long term, they will win."

---

[9] From "*UTC's Global Primer*", by William J. Holstein, *Chief Executive Magazine*, Volume 203, November 2004.

Maintaining Discipline and Balance

* * *

Despite its low corporate profile, United Technologies Corporation touches the lives of billions of people worldwide each day with its innovative products. The company has achieved an enviable position in the corporate world not only for it financial performance and ongoing success, but also for its disciplined management.

As brightly as anywhere else that discipline shines through in UTC's product creation processes. Even though the company is committed to the decentralization of its widely diverse operating units, certain principles and qualities remain consistent throughout the process of product creation enterprise-wide.

UTC's emphasis on continuous improvement—which permeates the organization—dates back to the late 1980s and the influence of the quality guru Ito. Formal product creation processes, which grew out of the quality culture, focus on multi-disciplinary teams, informed decision making, and risk reduction. The process also instills the courage to cancel projects that fail to live up to expectations.

Product creation throughout the company benefits from a centralized technical resource, the United Technologies Research Center, which often helps the company achieve technology synergies across business units. UTC also excels at portfolio management despite the challenges of maintaining numerous highly engineered product lines, most of which require years to develop and often have decades-long product life spans.

Operating discipline, cost reductions, fundamental innovation, and a culture of accountability have all helped sustain UTC as an industrial dynamo long-term. But perhaps most important of all to its success is UTC's unwavering commitment to investing in its human resources. Unlike so many others, UTC truly views its people as the company's most valuable asset and nurtures it accordingly.

Chapter 12

IBM

———— ✳ ✳ ✳ ————

Leveraging Research and Collaboration for Customer Innovation

In the early 1990s, IBM had what even long-time insiders still call a "near-death experience." With a reported record net loss of $8.1 billion in 1993, IBM appeared to be on its way to the corporate equivalent of the elephant graveyard. In 2001, just eight years later, IBM reported a net income of $7.7 billion, and its share price had increased nearly eight-fold.

What happened? Louis V. Gerstner happened, and the story of the almost miraculous turnaround of the IT behemoth is legendary. That story has been told countless times in business magazines, academic journals, and business school case studies, but perhaps never with more authority than in Gerstner's own 2002 book, *Who Says Elephants Can't Dance?* Dance indeed. Having once been on its way to oblivion, IBM today seems well on its way to recapturing its former glory as "the world's most admired company."

While Gerstner certainly deserves all of the accolades he's received for the turnaround, credit must also go to those who followed his leadership, took up his mantra, and executed his strategies. Credit for the continuation of IBM's success since Gerstner's retirement in 2002 must also go to Samuel J. Palmisano, IBM's current CEO, and his vision for the company's future, a vision unwavering in its focus on customers, research, and collaboration.

Those are the themes of the story that extends from Gerstner's entrance to the passing of the torch to Palmisano: an intense focus on understanding and meeting customer needs, a strong reliance on research to enable product creation, and an uncanny sense of how to balance collaboration with competition. One unifying aspect of all these themes also stands out: they all have been consistently supported by leadership from the very top of the organization.

Although these themes—focus on the customer and emphasis on research and collaboration—appear to permeate the entire enterprise, this chapter will limit its discussion of them as they apply to and inform product creation within the company's software group. But first a bit more background.

Bringing Big Blue out of the Red

✳ ✳ ✳

Reinvention is not a new phenomenon at IBM. The company has had many transformations. In its 100-year history as an American corporate icon, IBM has developed and successfully marketed generation after generation of "business machines" that support the creation and management of business information. Over that century, the company's products have evolved from computing scales to tabulators to work-time recorders to general purpose business computers to super computers to personal computers to software and business services. In its evolution, however, none of the company's transformations has been more dramatic in terms of sheer survival than the era of CEO Lou Gerstner.

After the company's heyday in the 1960s and 1970s, and pre-Gerstner, IBM had begun struggling to keep pace with its rapidly changing marketing environment. The Internet had already made its presence known, personal computing was quickly becoming the order of the day, and the ".com" mentality was just beginning to take root.

IBM appeared to have lost its edge, its focus, and to a considerable degree its brand identity. Its business units were siloed. Internal competition—for resources and business—sometimes seemed more intense than with external competitors. The company, far more often than not, displayed an extreme product focus, a lack of unity in direction, and an absence of a cohesive corporate strategy.

Upon his arrival in April 1993, Gerstner quickly set the company on a new course. In stark contrast to IBM's then prevailing corporate culture and strategy, Gerstner based his approach on a belief that the whole of IBM was worth far more than the sum of its parts. After only 30 days on the job, Gerstner introduced the concept of Team IBM and launched a reorganization and revitalization program that energized the employees enterprise wide. In some respects, his efforts have also had a lasting impact on the entire industry.

Prior to Gerstner's arrival, IBM's focus was highly product-centric, and product development was tightly confined within separate business units. Very little effort went into cross-product collaboration, much less cross-functional synergies. Gerstner brought a strong, almost obsessive, focus on understanding and fulfilling customers' needs.

He came to that task with considerable credentials having formerly been one of IBM's biggest customers as CEO of RJR Nabisco, Inc®. As Bob Biamonte, Vice President of IBM's Software Group World Wide Consumability and IBM User Technologies recalls, "Lou Gerstner challenged us to ensure that everything we did could be linked to creating value for the customer. His direction was clear, if you can't make that connection, stop what you're doing and do something else."

A cultural shift of major proportions. IBM's transformation probably would never have happened without an incredibly strong leader like Gerstner. Biamonte notes, "He launched the reintegration of IBM and got us focused on working as a seamless worldwide team to deliver value to our customers. When we started collaborating, it created an energy and spirit among the troops, both a top-down and bottoms-up effect. At the end of the day, it's all about culture, and shaping culture starts at the top." With this new focus Gerstner never wavered in his belief that the business units need to retain a strong P&L accountability, and this combination put IBM back on its feet.

Bill Woodworth, Director of IBM Quality Software Engineering, says that another tenet that Gerstner instilled, and one that still rings true, within the company is that "research is the foundation for IBM's competitive advantage." For Gerstner, R&D was a key enabler to future success for product innovation across the corporation.

Woodworth recalls, "Gerstner came up with a fundamental message, 'Research is our future.' Then he went to our colleagues in research with a clear charter to get their work accelerated and into products for our customers." Like his emphasis on a single value proposition and corporate strategy, Gerstner believed that a single corporate R&D unit was the glue that would hold product creation together across the various business units. Stuart Feldman, Vice President of Computer Science Research extends that idea a bit further. He says, "Research at IBM is embedded in the culture. The Research Division has full corporate responsibility and plays a vital corporate role alongside all the other corporate divisions."

Changing of the Guard

* * *

When Lou Gerstner decided to retire in 2002, instead of looking outward for a "knight in shining armor" as it had in 1993, IBM looked inward and tapped Samuel J. Palmisano for the top spot. Palmisano, who had joined IBM in 1973 at age 22, came to his new role with a reputation for being "true blue." That doesn't imply he was mired in the old ways, but rather that his experience and knowledge of the company, its people, and its products was vast and had prepared him to take the reins decisively. Where Lou Gerstner put IBM on its feet financially with his polices, Sam Palmisano is taking the company to the next level with continued focus on the customer and innovation.

Palmisano's actions speak louder than words. His first message to the industry painted a vision of the marketplace based on his own work sessions with the CEOs and CIOs of IBM's major customers, sessions, incidentally, that continue to this day. Palmisano's approach appears to be one of listening and communication. Starting with the customer's problem or challenges and then linking technology to the solution. Says Woodworth, "When Sam coined the phrase 'e-business on demand,' he was saying where we are going to take the industry. It's the grand vision at the top of our corporate strategy. All brands align in support of that."

Software Product Creation

✳ ✳ ✳

Currently IBM is organized into several product and service groups: Global Technology Services, Global Business Services, Software Group, Integrated Operations, Innovation and Technology, and Systems and Technology Group. While product creation synergies exist across all groups, the discussion here focuses primarily on the Software Group (SWG).

Leveraging its strong brand identity, the Software Group is organized into five business units as shown in Figure 12.1. It's important to note that although the group is organized into brands for P&L accountability and customer focus, it is matrix managed by functional groups to capitalize on cross-brand synergies.

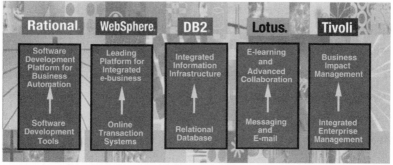

Figure 12.1 IBM Software Group Organization

IBM's Software Group, consistent with Gerstner's decade-old emphasis, adheres to a single vision and a single strategy. Across all of the brands, the group's off-the-shelf information technology products feature open architecture and compatibility with industry standards. Such a foundation helps the group ensure that its products embody the attributes that it has determined define a viable, successful software product. Those product attributes are scalable, flexible, extensible, resilient, virtualizable, manageable, and interoperable.

The overarching strategy of the Software Group is to focus on enabling integration through their middleware. Says Grace Lin, an IBM Distinguished Engineer and Global Sense and Respond Leader, "Every decision we make starts with the customer at the top of the pyramid. We want to create integrated solutions—connections to our middleware and systems. If there's a need, we help customers fill the gap."

CEO Sam Palmisano focuses on customer input by regularly running two different customer conferences, one for CEOs and one for CIOs, to get their high-level input. All senior executives meet with their peers around three geographies: the Americas, Europe, and Asia Pacific. These interactions are duplicated down the management chain and by using other forms of input and customer feedback. All brands have customer councils that focus on understanding customer needs and, perhaps more important, identifying issues or areas where customers think performance could be improved.

IBM's Product Development Process

* ✳ *

IBM has a well defined product development process called Integrated Product Development (IPD). A fairly classic stage-gate process, IPD is used enterprise-wide to guide all product creation and is the decision-making process that helps determine priorities and set requirements for product development.

IPD has three essential elements: clearly defined phases; periodic checkpoints for fact-based decision making; and cross-functional, team-based management. IPD consists of six distinct development phases: concept, plan, develop, qualify, launch, and lifecycle. The periodic decision checkpoints (DCPs), which occur between phases, are:

- Concept DCP, conducted at the end of the concept phase to provide initial project definition and project approval to proceed.

- Plan DCP, conducted at the end of the plan phase to establish overall project funding and basic functional commitments.

- Availability DCP, conducted at the end of the qualify phase to evaluate readiness for launch and approval for formal product announcement.

- End of life DCP, conducted at the end of the lifecycle phase to establish approval for withdrawal from market.

Each Software Group brand has it's own IPD organizational structure that includes representation from all brand stakeholders including brand marketing, development, product management, sales, support, quality, and finance.

- The top level cross-functional team is the Investment Review Board (IRB). This is usually chaired by the brand-level general manager. The IRB sets overall organizational strategy and priorities for investment.

- Within each brand, an Integrated-Portfolio Management Team (IPMT) manages the defined market segment within the brand. The IPMT provides focus on portfolio management and P&L and makes the DCP decisions. In some brands, the IRB and the IPMT are combined.

- The IPMT begins with the product business objectives as defined in the segment business plans, a very market-focused planning document. The end-to-end

IPD process establishes the development and research investments required, including support of the product in the marketplace, to support the product goals.

- Project Development Teams (PDTs), are the working groups that actually develop the products. At the beginning of concept development, the PDT is formed and includes a project leader and a virtual team with all stakeholders represented. The team, however, grows and shrinks over time as project requirements change.

The IPD process owner for IBM corporate is Nicholas Donofrio, Executive Vice President for Innovation & Technology. For the Software Group, the process owner is Carl Kessler, Vice President, WorldWide Development and Quality. Also integral to the process is Hira Advani, who serves as Director, SWG Community and System House, who assures that all process elements stay on track. He focuses on issue resolution and assures alignment of resources and activities from corporate headquarters through group and brand. An infrastructure called the "clearing house" helps to manage the interdependency and potential disconnects that exist among all aspects of and participants in product creation.

Because the IPD process is pervasive throughout IBM, staff at all levels of the company are trained in its principles and processes using a variety of methods from online self-study to formal classroom training sessions. Training is updated on a regular basis to ensure staff knowledge stays current with any process changes. Multiple communication channels are also used to "spread the word," including regular "lunch and learn" sessions and websites.

In 2005, the Software Group performed a "process check" to determine if IPD was delivering to its full potential. The exercise showed that IPD had become too bureaucratic, consumed too much overhead, included too many briefings, and did not focus clearly enough on the key business issues. In a serious re-tooling effort, IPD was streamlined to focus on the key essentials for decision making. This is not uncommon. In fact, one of the principles espoused in this book is that companies should anticipate and conduct periodic reviews of new product creation processes and refocus as necessary to deliver optimal results.

Research the Enabler and Integrator

* ✳ *

From the era of Gerstner forward to the present day, corporate research at IBM has played an integral role in new product creation. Stuart Feldman, Vice President of Computer Science Research, notes, "Research at IBM is embedded in the culture. You never hear the phrase 'tech transfer' at IBM. Each senior researcher has a partner in a development, marketing, or service division. It's a deep and formal relationship, and both parties are graded on it. You don't want to develop something that no one else wants."

At IBM, research also helps fill the gap that some IBM staff call the "white space" between business units in an emerging business area. The concept of white space ideas fits logically within the company's emphasis on P&L accountability for each business unit. Given that these businesses need clear scope definitions to minimize overlap, and that the information business is so dynamic and in constant flux, these "white space" issues and opportunities emerge quite frequently. Research helps identify them and supports the company in capturing the opportunities they may represent.

Although product focused research is the rule, the exception is also considered extremely important. This is evident in how the company funds its corporate research activities. The company allocates one-third of the research budget for base research and one-third to be used to match investments from business units, which make up the final third. Hence, the net funding profile is one-third discretionary and two-thirds focused on the specific identified needs of the business units.

Stuart Feldman says," The goal of research is to find and stay on the sweet spot so that research is neither overly focused on instant gratification nor so blue sky that it becomes disconnected from the business." He also notes that because the sweet spot is

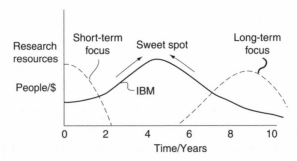

Figure 12.2 Time versus Resources/People in Research

relatively unstable, and research organizations tend to drift to either extreme, one of his jobs is to maintain the balance. "What you're doing should aim to be relevant in the foreseeable future," he says, "but not necessarily immediately."

According to David F. McQueeney, PhD, Vice President of Technology and Strategy and Chief Technology Officer of IBM Federal, IBM Software Group, "IBM tries to keep the sweet spot one to three years out with some tie to the present." He notes that some companies, GE for example, focus more on the short term, while others, such as Xerox PARC and the old AT&T Laboratories, work farther out. McQueeney is quick to note, however, that IBM is not rigid in defining the sweet spot, "Sometimes we allow and even encourage some projects to go way out in terms of time and risk/benefit." This "Sweet spot" concept is shown in figure 12.2.

Once a research project makes the transition into a "real" product in the field, the researcher often follows its deployment, sometimes in a support role during the product development process and sometimes by joining the business unit in a new role. It's not unusual to find former researchers filling high level strategic and corporate positions. In fact, the current corporate vice president in charge of strategy was formerly a vice president in the research group. Such cross-fertilization creates useful relationships that keep research in touch with the business and the business in touch with research.

The role of research is not static. It evolves just like the company itself and has transformed from a physics and engineering focus on semiconductors and hardware in the 1970s and 1980s to an emphasis on software—using math, computer science, and social science—to its currently evolving focus on service areas. Stuart Feldman notes that service is an area ripe for more research effort. "Service is something that gets consumed at the time of creation. A product is created in advance. The distinction is not trivial. A classic good is produced on specification in the hope that someone will buy it someday. A service is produced on the spot as someone wants it. There's a lot more that research can contribute to service development."

Looking outward. Research at IBM works diligently to look outward for ideas, inspiration, and direction. The group organizes formal forums and technical conferences to keep in touch with audiences as diverse as venture capitalists and academics to tap into new opportunities and to track emerging trends.

One of the company's most important touchstones in terms of research is the annual Global Technology Outlook (GTO). The GTO is essentially the research division's view of where technology is heading in both the near and long term. The GTO is not for "techies only". While a much more detailed and IBM-specific GTO is used for internal purposes, IBM also produces a more general version for use with collaborators, customers, and prospects.

The GTO has major impact, reaching as far as the executive suite. For each iteration of the GTO, IBM's CEO sets aside a full half-day session with the research group to learn firsthand the trends and forecasts for hardware and software development. Says Stuart Feldman, "It's almost unheard of for the CEO of any company to devote so much time to a pure technology and research meeting."

Collaboration and Cooperation

* ※ *

IBM has evolved, by design, to foster collaboration both internally and externally, clearly in support of the company's aim to satisfy customer needs and to support an open software system. Internally this means working in teams across business units and functions. Synergies come from both opportunistic and systematic approaches and from finding the right balance between dynamic and rigid approaches.

Eric Herness, Web Sphere Business Integration Chief Architect, WebSphere Development and IBM Distinguished Engineer, characterizes some of the less traditional approaches to collaboration as "ad hoc, social, opportunistic." These include blogs, team rooms, and other forms of social interaction technologies that encourage and support collaboration in all its forms.

Collaboration, of course, also means deriving innovation and technology from corporate research and other external research organizations. It can even mean working closely at times with competitors. Rob High, Service Oriented Architecture Foundation, Chief Architect, IBM Distinguished Engineer, Member, IBM Academy of Technology explains, "An odd mixture of relationships defines our professional lives, externally and internally. We rely heavily on collaboration in both arenas. For example, even though we compete with Oracle and SAP, we also collaborate with them on standards. It's in our mutual interest to make the end customers' solutions work. No enterprise stands alone as an island for very long."

Valuing People

* * *

Most organizations pay at least lip service to the value of their people. IBM delivers on the promise. Every IBM employee has an individual development plan (IDP) that concentrates on who they are, what they can do, what they would like to do, and finding the right path to develop their full potential, including the resources they need.

Two programs in particular—the Technical Resource (TR) Program and the Executive Resource (ER) Program—begin relatively early in a staff member's tenure to identify those who have the talent and temperament to perhaps someday run the company in either a technical or executive management leadership position or both. Line management begins looking early for those who show promise for becoming IBM's business executives or distinguished engineers and fellows of the future. Those identified receive focused mentoring and attention up and down the line management chain. Top talent emerges at the upper end with the top people in ER becoming senior executives and the top people in TR becoming either IBM Distinguished Engineers or Corporate Fellows.

Annually, Sam Palmisano sits with his group executives to review the top tier of managerial and technical talent. All reviews focus on the development needs of the individuals and how to ensure they have the resources, experiences, and guidance they need to excel. Rather than hoarding talent in one area, IBM seeks to give its best and brightest the broadest preparation possible for personal success and for making valuable contributions to the company.

At IBM, there's also a rich heritage of mentoring. Every senior technical leader or manager is encouraged to serve as a mentor for more junior employees. While mentoring is encouraged for everyone, it is required for those in the ER and TR programs. Each participant must not only have a mentor, but must also serve as a mentor for others.

Two other formal programs help support long-term career development of young employees. A promising individual with three to five years experience can "shadow" an IBM Distinguished Engineer (DE) or business executive for up to a month. The person literally follows the technical or business executive, participating in whatever the executive works on, joining in patent applications, and meeting with customers. This provides the employee a firsthand experience to observe and learn about the technical or business executive's daily life and what makes him or her special.

Another program for promising young employees is called "Extreme Blue/BizTech," The Extreme Blue program was developed to attract top talent from universities through a summer internship program. Participants in Extreme Blue are assigned challenging projects and receive executive attention and mentoring throughout the program. At the end of the term, all teams present their results to a group of IBM executives. As the program became a powerful IBM recruiting tool for top talent, IBM expanded it to young employees with five or fewer years of experience to improve retention rates. The participants spend about 20 percent of their time for ten months to a year working on a high profile project. Four times a year, all participants meet with other team members, culminating with presentations to the senior executives who sponsor the projects.

A Well Chosen Path

* * *

IBM has made enormous strides in its comeback from the tough years of only a decade ago. Today, in typical IBM fashion, the company appears to be constantly reinventing itself in terms of products and services with its culture characterized by innovation and collaboration. While IBM's product development process appears traditional in structure, its effectiveness derives to a large degree from the integration of all stakeholders in the development decision-making process. Perhaps even more, it comes from the company's ability to leverage its corporate research function to drive innovation and create new business opportunities. At IBM, this is a highly developed capability rarely seen to such a degree in organizations of any size or in any industry.

INDEX

ABOUT THE AUTHORS

Ron Kerber is an independent consultant and former Executive Vice President and Chief Technology Officer of Whirlpool Corporation. For a decade, he was a member of Whirlpool's Executive Committee and held line responsibility for the company's global product development and procurement programs. He also managed global business lines with annual sales totaling more than $1.2 billion and a total organization of over 6000 employees.

Prior to joining Whirlpool, Dr Kerber served as Corporate Vice President, Technology and Business Development, and as a member of the Executive Committee, at McDonnell Douglas Corporation. In holding the senior engineering position he had corporate responsibility for product development, strategic planning, business development and procurement.

From 1985 to 1988, he served as Deputy Undersecretary of Defense for Research and Advanced Technology. In this position he was responsible for all DOD research and product development up to specific weapon system development. His office also managed DOD's Very High Speed Computer, Millimeter wave Microwave Computer and Software Programs and the Balanced Technology Initiative. He served on NATO and other ally committees responsible for technology cooperation. Prior to that he was a Staff Member at The Defense Advanced Research Projects Agency.

Upon leaving graduate school, Kerber joined Michigan State University where he rose through the ranks to Professor of Electrical Engineering and Mechanical Engineering and Associate Engineering Dean of Graduate Studies and Research. While on academic leave, he was a Member of the Staff of The Aerospace Corporation responsible for laser and materials research.

Since retiring from Whirlpool in 2001, Dr Kerber has devoted himself to a variety of entrepreneurial and pro bono activities. In addition to advising clients about global sourcing, product development, and manufacturing strategies, he is Co-Founder of Dominion Development Company in Charlottesville, Virginia. He is a visiting professor and Batten Fellow at the Darden Business School of the University of Virginia, a member of the Board of Anser Corporation and a member of the Defense Department's Defense Science Board.

Dr Kerber holds a Bachelor of Science degree from Purdue University and Master's and Doctorate degrees in engineering science from The California Institute of Technology. He has over seventy technical and policy publications and numerous awards including the Secretary of Defense Medal for Outstanding Public Service.

Tim Laseter serves on the faculty of the Darden Graduate School of Business Administration at the University of Virginia, one of the top-ranked business schools in the country. He joined the faculty in 2002 to teach operations in the first year core program plus an elective covering operations strategy. In addition to his duties in the core MBA program, Laseter teaches both open enrollment and custom courses for Darden's Executive Education programs. Recognized by the Dean's office for ranking in the top 20 percent among Darden faculty and nominated for the outstanding faculty award, Laseter also serves as faculty advisor for the Consulting Club and leads the case competition cosponsored by PepsiCo and the General Management and Operations Club.

Prior to joining the Darden faculty, Laseter served as an advisor to senior executives as a partner with Booz Allen Hamilton. Founder of the firm's global network of sourcing practitioners he worked with clients in a variety of industries including automotive, computers, defense, energy, media, and telecommunications. During his fifteen years with the firm, he gained a global perspective by transferring among a variety of Booz Allen offices, working out of Cleveland, London, New York, and McLean. Engagements addressed a wide range of issues including overall business strategy, organization, supply chain management, product development, sourcing, and related topics of operations strategy and spanned the globe including the United States, Europe, South America, and Asia.

A prolific writer for business executives, he is the author of *Balanced Sourcing: Cooperation & Competition in Supplier Relationships* (Jossey-Bass, 1998). Additionally he has authored or co-authored dozens of practitioner articles and book chapters plus numerous academic cases and peer-reviewed articles. His research has been cited in a range of publications including *The Wall Street Journal, The New Yorker*, and the *Progressive Grocer*. Laseter serves as a contributing editor for *strategy+business* and authors a recurring column on "Operating Strategies."

Prior to joining Booz Allen, Laseter worked at Siecor—at the time, a fiber optics joint venture between Siemens and Corning. He rotated through several positions starting as a business analyst in the Operations Controller's group then moving to a position as a night shift supervisor in the largest plant and finally as the head of the

Quality Assurance department. Earlier in his career he worked as a management consultant with Arthur Andersen and the McLean Group.

Laseter earned a Bachelor of Science degree in Industrial Management from the Georgia Institute of Technology with high honors. He holds a Masters in Business Administration and a doctorate in Operations Management from the Darden Graduate Business School at the University of Virginia and was a recipient of the Faculty Award for Academic Excellence.

Max Russell is a professional business writer who has authored and collaborated on numerous books, articles, and corporate publications on subjects as diverse as corporate governance, professional ethics, corporate codes of conduct, environmental and social responsibility, corporate reporting, new product development, business process redesign, post-merger integration, risk management, and health care reform. As a book writer, he most recently co-authored with Scott C. Newquist the well-received Putting Investors First: Real Solutions for Better Corporate Governance (Bloomberg Press, 2003). Mr. Russell holds an MBA from Northwestern University's Kellogg School of Management.